Combinatorial Problems in Mathematical Competitions

Mathematical Olympiad Series

ISSN: 1793-8570

Series Editors: Lee Peng Yee *(Nanyang Technological University, Singapore)*
Xiong Bin *(East China Normal University, China)*

Published

Vol. 1   A First Step to Mathematical Olympiad Problems
by Derek Holton (University of Otago, New Zealand)

Vol. 2   Problems of Number Theory in Mathematical Competitions
by Yu Hong-Bing (Suzhou University, China)
translated by Lin Lei (East China Normal University, China)

Vol. 3   Graph Theory
by Xiong Bin (East China Normal University, China) &
Zheng Zhongyi (High School Attached to Fudan University, China)
translated by Liu Ruifang, Zhai Mingqing & Lin Yuanqing
(East China Normal University, China)

Vol. 4   Combinatorial Problems in Mathematical Competitions
by Yao Zhang (Hunan Normal University, P. R. China)

Vol. 5   Selected Problems of the Vietnamese Olympiad (1962–2009)
by Le Hai Chau (Ministry of Education and Training, Vietnam)
& Le Hai Khoi (Nanyang Technology University, Singapore)

Vol. 6   Lecture Notes on Mathematical Olympiad Courses:
For Junior Section (In 2 Volumes)
by Xu Jiagu

Yao Zhang

Hunan Normal University, P R China

Vol. 4 | Mathematical Olympiad Series

Combinatorial Problems in Mathematical Competitions

East China Normal University Press

World Scientific

Published by

East China Normal University Press
3663 North Zhongshan Road
Shanghai 200062
China

and

World Scientific Publishing Co. Pte. Ltd.
5 Toh Tuck Link, Singapore 596224
USA office: 27 Warren Street, Suite 401-402, Hackensack, NJ 07601
UK office: 57 Shelton Street, Covent Garden, London WC2H 9HE

British Library Cataloguing-in-Publication Data
A catalogue record for this book is available from the British Library.

COMBINATORIAL PROBLEMS IN MATHEMATICAL COMPETITIONS

ISBN-13 978-981-283-949-7 (pbk)
ISBN-10 981-283-949-6 (pbk)

Printed in Singapore.

Introduction

This book consists of three parts: fundamental knowledge, basic methods and typical problems. These three parts introduce the fundamental knowledge of solving combinatorial problems, the important solutions to combinatorial problems and some typical problems with often-used solutions in the high school mathematical competition respectively.

In each chapter there are necessary examples and exercises with solutions. These examples and exercises are of the same level of difficulty as the China Mathematical League Competitions which are selected from mathematical competitions at home and abroad in recent years. Some test questions are created by the author himself and a few easy questions in China Mathematical Olympiad (CMO) and IMO are also included. In this book, the author pay attention to leading readers to explore, analyze and summarize the ideas and methods of solving combinatorial problems. The readers' mathematical concepts and abilities will be improved remarkably after acquiring knowledge from this book.

Preface

Combinatorial Mathematics has a long history. Thousands of years ago, some simple and interesting combinatorial problems started to be involved in the ancient Chinese masterpieces such as *He Tu* and *Luo Shu*. In the recent 20 years, owing to the rapid development of knowledge such as Computer Science, Coding Theory, Programming Theory, Digital Communication and Experimental Designing, a series of theoretical and practical problems need to be solved by discrete mathematics. Moreover, the questions which were raised by the logic of combinatorial mathematics itself and other branches of mathematics, has made the research about combinatorial mathematics to flourish and very rich. The problem-solving techniques and methods are daedal, transforming this ancient mathematical idea into a rigorous mathematical subject.

The combinatorial questions in the mathematical competitions are always straightforward, but the people who solve these problems should be endowed with the power of acute observation, rich imagination and necessarily skills. There are no fixed methods that can be followed, and various questions of different level of difficulties are very rich. Hence, combinatorial questions have been included in different levels of intelligence test and mathematical competition.

This book focus on the combinatorial problems in the mathematics competition, and consists of three parts: fundamental knowledge, basic methods and typical problems. This book is a translation from the author's book of the same title (in Chinese), with a few amendments and supplements. It not only emphasizes the fundamental knowledge of solving combinatorial problems in mathematical competitions, but also introduces to the readers the important solutions to combinatorial problems and some typical problems with solutions that are often used.

In the choice of examples and exercises, except the questions which strikingly newer have been selected as far as possible, but the novel test questions which are the same difficult level as the China Mathematical League Competitions in the mathematical competition at home and abroad in recent yeas have been chosen deliberately. Some test questions created by the author himself and a few easy questions in China Mathematical Olympiad (CMO) and IMO are also included. In this book, the author pays attention to leading the readers to explore, analyze and summarize the ideas and methods of solving combinatorial problems. The readers' mathematical concepts and abilities will be improved remarkably after acquiring the knowledge from this book.

Contents

1. 1   Two Basic Counting Principles

The Addition Principle

If there are n_1 different objects in the first set, n_2 objects in the second set, ..., and n_m objects in the mth set, and if the different sets are disjoint, then the number of ways to select an object from one of the m sets is $n_1 + n_2 + \cdots + n_m$.

The Multiplication principle

Suppose a procedure can be broken into m successive (ordered) stages, with n_1 outcomes in the first stage, n_2 outcomes in the second stage, ..., and n_m outcomes in the mth stage. If the composite outcomes are all distinct, then the total procedure has $n_1 n_2 \cdots n_m$ different composite outcomes.

Example 1   How many ways are there to choose 4 distinct positive integer numbers x_1, x_2, x_3, x_4 from the set $S = \{1, 2, \ldots, 499, 500\}$ such that x_1, x_2, x_3, x_4 is an increasing geometric sequence and its common ratio is a positive integer number?

Solution   Let $a_1, a_1q, a_1q^2, a_1q^3$ $(a_1, q \in \mathbf{N}_+, q \geqslant 2)$ be the four numbers which are chosen by us, then $a_1 q^3 \leqslant 500$, $q \leqslant \sqrt[3]{\dfrac{500}{a_1}} \leqslant \sqrt[3]{500}$.

Hence $2 \leqslant q \leqslant 7$, and $1 \leqslant a_1 \leqslant \left[\dfrac{500}{q^3}\right]$, that is the number of the geometric sequences with the common radio q is $\left[\dfrac{500}{q^3}\right]$. By the addition principle, the number of the geometric sequences satisfying the conditions is

$$\sum_{q=2}^{7} \left[\frac{500}{q^3}\right] = 62 + 18 + 7 + 4 + 2 + 1 = 94.$$

So the answer to the question is 94.

Example 2 How many 4-digit odd numbers with distinct digits are there?

Solution A 4-digit number is an ordered arrangement of 4 digits (leading zeros not allowed). Since the numbers we want to count are odd, the unit digit can be any one of 1, 3, 5, 7, 9. The tens digit and the hundreds digit can be any one of 0, 1, ..., 9, while the thousands digit can be any one of 1, 2, ..., 9. Thus there are 5 choices for the unit digit. Since the digits are distinct, we have 8 choices for the thousands digit, whatever the choice of the unit digit is. Then there are 8 choices for the hundreds digit, whatever the first 2 choices are, and 7 choices for the tens digit, whatever the first 3 choices are. Thus by the multiplication principle, the number of 4-digit odd numbers with distinct digits is equal to $5 \times 8 \times 8 \times 7 = 2240$.

1.2   Permutation Without Repetition and Combination Without Repetition

Permutation An ordered arrangement of n distinct objects taking m ($m \leqslant n$) distinct objects at a time is called a permutation of n distinct objects taking m distinct objects at a time. Since the objects is not repeated, the permutation is also called the permutation without repetition, and the number of "permutation of n distinct objects taking m distinct objects" is denoted by P_m^n or A_m^n [*] , then

$$P_m^n = n(n-1)(n-2)\cdots(n-m+1) = \frac{n!}{(n-m)!},$$

---

[*] $P_m^n (A_m^n)$ is also written as $P_n^m (A_n^m)$ in some countries.

where $m \leqslant n$, and there is a convention $0! = 1$.

Especially, when $m = n$, the permutation of n distinct objects taken n distinct objects is called all permutation of n distinct objects. The number of all permutation of n distinct objects is equal to

$$P_n^n = n(n-1)(n-2)\cdots 2 \cdot 1 = n!$$

Combination   An unordered selection of n distinct objects taking m $(m \leqslant n)$ distinct objects at a time is called a combination of n distinct objects taking m distinct objects at a time. Since the objects is not repeated, a combination of n distinct objects taking m distinct objects is also called a combination without repetition. The number of "combination of n distinct objects taking m distinct objects" is denoted by $\binom{n}{m}$, then

$$\binom{n}{m} = \frac{P_m^n}{m!} = \frac{n(n-1)(n-2)\cdots(n-m+1)}{m!} = \frac{n!}{m!(n-m)!}.$$

Example 3   How many 5-digit numbers greater than 21300 are there such that their digits are distinct integers taken from $\{1, 2, 3, 4, 5\}$.

Solution I   We divide these 5-digit numbers satisfying the required conditions into 3 types:

The number of 5-digit number whose ten thousands digit may be any one of 3, 4 or 5 is equal to $P_1^3 P_4^4$.

The number of 5-digit number whose ten thousands digit be 2 and thousands digit be any one of 3, 4 or 5 is equal to $P_1^3 P_3^3$.

The number of 5-digit number of ten thousands digit be 2, and thousands digit be 1 is equal to P_3^3.

By the addition principle, the number of 5-digit numbers satisfying the required conditions is equal to $P_1^3 P_4^4 + P_1^3 P_3^3 + P_3^3 = 96$.

Solution II   Since the number of 5-digit numbers with distinct digits taken from 1, 2, 3, 4, 5 equals P_5^5, and there are only P_4^4 numbers (their ten thousands digit are equal to 1) not exceeding 21300. Hence the number of 5-digit numbers satisfying the required conditions equals $P_5^5 - P_4^4 = 96$.

1.3   Repeated Permutation and Repeated Combination

Repeated Permutation   An ordered arrangement of n distinct objects taking m objects at a time (each object may has a finite repetition number) is called a repeated permutation of n distinct objects taken m objects at a time. The number of this repeated permutation is equal to n^m.

This conclusion could be proved easily by the multiplication principle.

Repeated Combination   An unordered selection of n distinct objects taking m objects (each object may has a finite repletion number) is called a repeated combination. The number of this repeated combination is equal to $\binom{n+m-1}{m}$.

Proof   Denote the n distinct objects by 1, 2, $\cdots$, n. Then repeated combination of n distinct objects taken m objects has the following form: $\{i_1, i_2, \ldots, i_m\}$ $(1 \leqslant i_1 \leqslant i_2 \leqslant \cdots \leqslant i_m \leqslant n)$. Since the selections could be repeated, so that the equality holds. Set $j_1 = i_1, j_2 = i_2 + 1, \cdots, j_m = i_m + (m-1)$, then $1 \leqslant j_1 < j_2 < \cdots < j_m \leqslant n+m-1$, and the $\{j_1, j_2, \ldots, j_m\}$ is just the combination without repetition of $n+m-1$ distinct objects: 1, 2, $\ldots$, $n+m-1$ taken m distinct objects.

Hence the number of the required repeated combination equals $\binom{n+m-1}{m}$.

All Permutation of Incomplete Distinct Objects   Suppose that n objects consist of k distinct objects $a_1, a_2, \ldots, a_k$ with repetition numbers $n_1, n_2, \ldots, n_m (n_1 + n_2 + \cdots + n_m = n)$ respectively, the all permutation of these n objects is called the all permutations of the incomplete distinct objects. We denote the number of all such permutation by $\begin{pmatrix} n \\ n_1, n_2, \ldots, n_k \end{pmatrix}$, then $\begin{pmatrix} n \\ n_1, n_2, \ldots, n_k \end{pmatrix} =$ $\dfrac{n!}{n_1! n_2! \cdots n_k!}$.

Proof Let f denote the number of the all permutation satisfying the conditions. If we exchange the same objects in each kind for the mutually distinct objects and rearrange them, then we get $n_1! n_2! \cdots n_k!$ all permutations of n distinct objects. By the multiplication principle, the number of the all permutation of n distinct objects is equal to $f \cdot n_1!$ $n_2! \cdots n_k!$. But the number of all permutation of n distinct objects is equal to $n!$. Hence $f \cdot n_1! n_2! \cdots n_k! = n!$. Thus

$$f = \begin{pmatrix} n \\ n_1, n_2, \ldots, n_k \end{pmatrix} = \frac{n!}{n_1! n_2! \cdots n_k!}.$$

Multiple Combination Let's classify n distinct objects into k ($k \leqslant n$) distinct kinds, such that there are n_i objects in i^{th} kind ($i = 1, 2, \ldots, k$, $n_1 + n_2 + \cdots + n_k = n$). Then the number of the classify ways is equal to $\begin{pmatrix} n \\ n_1, n_2, \ldots, n_k \end{pmatrix} = \dfrac{n!}{n_1! n_2! \cdots n_k!}$.

Proof Since the number of ways of the n distinct objects taken n_1 distinct objects is equal to $\begin{pmatrix} n \\ n_1 \end{pmatrix}$. Then, the number of ways taking n_2 distinct objects from the residual $n - n_1$ distinct objects is equal to $\begin{pmatrix} n - n_1 \\ n_2 \end{pmatrix}$. If we continue like this and invoke the multiplication principle, we find that the number of distinct partitioned kinds equals

$$\begin{pmatrix} n \\ n_1 \end{pmatrix} \begin{pmatrix} n - n_1 \\ n_2 \end{pmatrix} \cdots \begin{pmatrix} n - n_1 - n_2 - \cdots - n_{k-1} \\ n_k \end{pmatrix}$$

$$= \frac{n!}{n_1! (n - n_1)!} \cdot \frac{(n - n_1)!}{n_2! (n - n_1 - n_2)!} \cdots \frac{(n - n_1 - \cdots - n_{k-1})!}{n_k! (n - n_1 - \cdots - n_k)!}$$

$$= \frac{n!}{n_1! n_2! \cdots n_k!}.$$

Remark The counting formulas of all permutation of incomplete distinct objects and multiple combination are the same, but their significance is different. We may prove the counting formula of the multiple combination by applying the same method of proving the formula of all permutation of incomplete distinct objects.

Example 4   In how many ways can one chose 10 paper currencies from the bank and the volumes of these paper currencies are 1 Jiao, 5 Jiao, 1 Yuan, 5 Yuan, 10 Yuan 50 Yuan and 100 Yuan respectively? (**Remark:** The Jiao and Yuan are the units of money in China.)

Solution   We are asked to count the repeated combinational number of ways to take 10 paper currencies from 7 distinct paper currencies. Using the formula of repeated combinatorial number, we get that the number of required distinct ways equals

$$\binom{7+10-1}{10} = \binom{16}{6} = \frac{16 \times 15 \times 14 \times 13 \times 12 \times 11}{1 \times 2 \times 3 \times 4 \times 5 \times 6} = 8008.$$

Example 5   Suppose that 3 red-flags, 4 blue-flags and 2 yellow-flags are placed on 9 numbered flagpoles in order (every flagpole hangs just one flag). How many distinct symbols consist of these flags are there?

Solution   Using the formula of all permutation number of incomplete distinct objects, we get that the number of distinct symbols equals $\binom{9}{3,\ 4,\ 2} = \frac{9!}{3!4!2!} = 1260.$

Example 6   How many are there to choose 3 pairs of players for the doubles from n ($\geqslant 6$) players.

Solution I   The number of taking 6 players from n distinct players equals $\binom{n}{6}$. The 6 players is classified into three groups such that each group contains exactly 2 players and the number of methods equals $\binom{6}{2,\ 2,\ 2}$, but the 3 groups are unordered, so the number required ways is equal to

$$\frac{\binom{n}{6}\binom{6}{2,\ 2,\ 2}}{3!} = \frac{n!}{6!(n-6)!} \cdot \frac{6!}{2!2!2!} \cdot \frac{1}{3!} = \frac{n!}{48(n-6)!}.$$

Solution II   The number of ways of taking 6 players from n distinct players equals $\binom{n}{6}$. Within the 6 players, there are $\binom{6}{2}$ ways to choose 2 players, and within the remaining 4 players there are $\binom{4}{2}$ ways to choose 2 players. Finally, there are $\binom{2}{2}$ ways to choose 2 players with in the remaining 2 players. But the 3 pairs are unordered, so that the number of required ways is equal to

$$\frac{\binom{n}{6}\binom{6}{2}\binom{4}{2}\binom{2}{2}}{3!} = \frac{n!}{48 \cdot (n-6)!}.$$

Remark   If we change this problem to the following problem "How many are there to choose 3 pair of player who serve as top seed players, second seed players and third seed players respectively, from n ($\geqslant 6$) players?" Then the number of different ways equals

$$\binom{n}{6}\binom{6}{2}\binom{4}{2}\binom{2}{2} = \frac{n!}{8 \cdot (n-6)!}.$$

Since these 3 pair players are ordered, it is not divided by 3!

1.4   Circular Permutation of Distinct Elements and Number of Necklaces

Circular Permutation of Distinct Elements   If we arrange the n distinct objects in a circle, then this permutation is called a circular permutation of n distinct objects. The number of circular permutation of n distinct objects equals $\dfrac{P_n^n}{n} = (n-1)!$.

Proof   Since n linear permutations $A_1 A_2 \cdots A_{n-1} A_n$, $A_2 A_3 \cdots A_n A_1$, $\ldots$, $A_n A_1 \cdots A_{n-2} A_{n-1}$ give rise to the same circular permutation and there are P_n^n linear permutations. Thus the number of circular permutations of n distinct objects equals $\dfrac{P_n^n}{n} = (n-1)!$.

Number of Necklace

Suppose that a necklace consists of n distinct beads which are arranged in circle, then the number of distinct necklaces is 1 (if $n = 1$ or 2) or $\dfrac{1}{2} \cdot (n-1)!$ (if $n \geqslant 3$).

Proof If $n = 1$ or 2, then the number of necklace is 1. Assume that $n \geqslant 3$. Since a necklace can be rotated or turned over without any change, the number of necklaces is one-half of the number of circular permutation of n distinct objects, i. e. $\dfrac{1}{2} \cdot (n-1)!$.

Example 7   How many ways are there to arrange 6 girls and 15 boys to dance in a circle such that there are at least two boys between any two adjacent girls?

Solution   First, for every girl, we regard two boys as her dancing partner such that one is at the left of this girl and another is at the right. Since 6 girls are distinct, we can select 12 boys from 15 boys in P_{12}^{15} ways. Next, every girl and her two dancing partners are considered as a group, each of residual $15 - 12 = 3$ boys are also considered as a group. Thus the total of groups is 9, and we can arrange them in a circle in $(9 - 1)! = 8!$ ways. By the multiplication principle, the number of permutations satisfying the conditions equals $P_{12}^{15} \cdot 8! = \dfrac{15! \cdot 8!}{3!}$.

1. 5   The Number of Solutions of the First Degree Indefinite Equation

The number of Solutions of The Indefinite Equation   The number of nonnegative integer solutions $(x_1, x_2, \ldots, x_m)$ of the indefinite equation $x_1 + x_2 + \cdots + x_m = n$ ($m, n \in \mathbf{N}_+$) is equal to $\dbinom{n+m-1}{m-1} =$

$$\binom{n+m-1}{n}.$$

Proof We consider that each nonnegative integer solution $(x_1, x_2, \ldots, x_m)$ of the equation $x_1 + x_2 + \cdots + x_m = n$ $(m, n \in \mathbf{N}_+)$ corresponds to a permutation of n circles "O" and $m-1$ bars "|":

$$\underbrace{OO\cdots O}_{x_1} \mid \underbrace{OO\cdots O}_{x_2} \mid \cdots \mid \underbrace{OO\cdots O}_{x_m}$$

Where x_1 is the number of circles "O" at the left of first bar "|", x_{i+1} is the number of circles "O" between the ith bars "|" and the $(i+1)^{\text{th}}$ bars "|", $\cdots$, x_m is the number of circles "O" at the right of the $(m-1)^{\text{th}}$ bar " | ". Since the correspondence is an one-to-one correspondence, the number of nonnegative integer solutions $(x_1, x_2, \ldots, x_m)$ of the indefinite equation $x_1 + x_2 + \cdots + x_m = n$ $(m, n \in \mathbf{N}_+)$ equals the number of the permutations of n circles "O" and $(m-1)$ bars "|", i. e. $\binom{n-m+1}{m-1} = \binom{n+m-1}{n}$.

Remark The number of nonnegative integer solutions $(x_1, x_2, \ldots, x_m)$ of the indefinite equation $x_1 + x_2 + \cdots + x_m = n$ $(m, n \in \mathbf{N}_+)$ is equal to the number of the repeated combinations from n distinct objects taken m objects (each object may has a finite repletion number).

Corollary The number of positive integer solutions $(x_1, x_2, \ldots, x_m)$ of the indefinite equation $x_1 + x_2 + \cdots + x_m = n$ $(m, n \in \mathbf{N}_+, n \geqslant m)$ equals $\binom{n-1}{m-1}$.

Proof Setting $y_i = x_i - 1$ $(i = 1, 2, \cdots, m)$, we get $y_1 + y_2 + \cdots + y_m = n - m$. Thus the number of positive integer solutions $(x_1, x_2, \ldots, x_m)$ of the indefinite equation $x_1 + x_2 + \cdots + x_m = n$ $(m, n \in \mathbf{N}_+, n \geqslant m)$ equals number of nonnegative integer solutions $(y_1, y_2, \ldots, y_m)$ of the indefinite equation $y_1 + y_2 + \cdots + y_m = n - m$, i. e.

$$\binom{(n-m)+m-1}{m-1} = \binom{n-1}{m-1}.$$

Applying above formula, we give another solution of example 7.

Second Solution of Example 7   Suppose that 15 boys are divided into 6 groups such that the leader of every group is a girl and there are at least two boys in every group. Denote the number of the boys in every group by x_1, x_2, ... , x_6 respectively, then

$$x_1 + x_2 + \cdots + x_6 = 15 \ (x_i \in \mathbf{N}_+ \text{ and } x_i \geqslant 2, i = 1, 2, \ldots, 6).$$
①

Setting $y_i = x_i - 2 \ (i = 1, 2, \ldots, 6)$, we get

$$y_1 + y_2 + \cdots + y_6 = 3 \ (y_i \in \mathbf{Z} \text{ and } y_i \geqslant 0, i = 1, 2, \ldots, 6).$$
②

Thus the number of the integer solutions of ① is equal to the number of the nonnegative integer solutions of ②, i. e. $\dbinom{3+6-1}{6-1} = \dbinom{8}{5} = \dbinom{8}{3}$. Hence the 15 boys are divided into 6 groups such that there are at least two boys in every group in $\dbinom{8}{3}$ ways. We arrange the 6 groups in a circle in $(6-1)! = 5!$ ways. (The leader of every group is a girl and her position is definite.) 15 boys stand in this circle in 15! ways. By the multiplication principle, we get that the number of the permutations satisfying requirement equals $\dbinom{8}{3} \cdot 5! \cdot 15! = \dfrac{8! \cdot 15!}{3!}$.

Example 8   How many 3-digit integers are there such that the sum of digits of each integer is equal to 11?

Solution   We denote the hundred digit, ten digit and unit digit by the x_1, x_2, x_3 respectively, then

$$x_1 + x_2 + x_3 = 11 \ (x_1 \geqslant 1, x_2 \geqslant 0, x_3 \geqslant 0).$$
①

Setting $y_1 = x_1 - 1$, $y_2 = x_2$, $y_3 = x_3$, we get

$$y_1 + y_2 + y_3 = 10 \ (y_1 \geqslant 0, y_2 \geqslant 0, y_3 \geqslant 0).$$
②

Thus the number of integer solutions of ① equals the number of nonnegative integer solutions of ②, i.e. $\binom{10+3-1}{3-1} = \binom{12}{2}$. But the

3-digit numbers cannot consist of the following 5 integer solutions of ①: (11, 0, 0), (10, 1, 0), (10, 0, 1), (1, 10, 0), (1, 0, 10), so that

the number of the 3-digit numbers satisfying conditions equals $\binom{12}{2} -$

$5 = 61$.

1.6   The Inclusion – Exclusion Principle

The Inclusion – Exclusion Principle   Let A_1, A_2, $\ldots$, A_n be n finite sets. We denote the number of elements of A_i by $|A_i|$ ($i = 1$, 2, $\ldots$, n). Then

$$|A_1 \cup A_2 \cup \cdots \cup A_n| = \sum_{i=1}^{n} |A_i| - \sum_{1 \leqslant i < j \leqslant n} |A_i \cap A_j|$$

$$+ \sum_{1 \leqslant i < j < k \leqslant n} |A_i \cap A_j \cap A_k| - \cdots \qquad ①$$

$$+ (-1)^{n-1} |A_1 \cap A_2 \cap \cdots \cap A_n|.$$

Proof   For any $x \in A_1 \cup A_2 \cup \cdots \cup A_n$, we show that x contributes the same count to each side of ①.

Since x belongs to at least one set of A_1, A_2, $\ldots$, A_n. Without loss of generality, let x belongs to A_1, A_2, $\ldots$, A_k and not belong to other sets. In this case, x is counted one time in $A_1 \cup A_2 \cup \cdots \cup A_n$.

But at the right side of ①, x is counted: $\binom{k}{1}$ times in $\sum_{i=1}^{n} |A_i|$, $\binom{k}{2}$ times

in $\sum_{1 \leqslant i < j \leqslant n} |A_i \cap A_j|$, $\cdots$, $\binom{k}{3}$ times in $\sum_{1 \leqslant i < j < k \leqslant n} |A_i \cap A_j \cap A_k|$, $\cdots$.

Consequently, at the right side of ①, x is counted

$$\binom{k}{1} - \binom{k}{2} + \binom{k}{3} - \cdots + (-1)^{k-1} \binom{k}{k}$$

$$= \binom{k}{0} - \left(\binom{k}{0} - \binom{k}{1} + \binom{k}{2} - \binom{k}{3} - \cdots + (-1)^k \binom{k}{k} \right)$$

$$= 1 - (1-1)^k = 1 \text{ time.}$$

Obviously for any $x \notin A_1 \cup A_2 \cup \cdots \cup A_n$, at the left side and right side of ① x is counted zero time.

Therefore, for any element x, at the two sides of ① x is counted the same time and the equality ① is verified.

Remark   The above method to prove equality ① is called the contributed method.

Successive Sweep Principle (Sieve Formula)   Let S be a finite set, $A_i \subset S(i = 1, 2, \ldots, n)$ and denote the complement of A_i in S by $\overline{A_i}$ $(i = 1, 2, \ldots, n)$, then

$$
\begin{aligned}
| \overline{A_1} \cap \overline{A_2} \cap \cdots \cap \overline{A_n} | &= | S | - | A_1 \cup A_2 \cup \cdots \cup A_n | \\
&= | S | - \sum_{i=1}^{n} | A_i | + \sum_{1 \leqslant i < j \leqslant n} | A_i \cap A_j | \\
&\quad - \cdots + (-1)^n | A_1 \cap A_2 \cap \cdots \cap A_n |.
\end{aligned}
$$

②

Proof   Since

$$
| \overline{A_1 \cup A_2 \cup \cdots \cup A_n} | = | S | - | A_1 \cup A_2 \cup \cdots \cup A_n | \qquad ③
$$

By De Morgan's Laws, we obtain

$$
| \overline{A_1 \cup A_2 \cup \cdots \cup A_n} | = | \overline{A_1} \cap \overline{A_2} \cap \cdots \cap \overline{A_n} | . \qquad ④
$$

Combining ③ and ④ with ①, we deduce the equality ② immediately.

Example 9   Determine the number of positive integers less than 1000 which are divisible by neither 7 nor 5.

Solution   Let $S = \{1, 2, \ldots, 999\}$, $A_i = \{k \mid k \in S, k$ is divisible by $i\}$, $(i = 5$ or $7)$. Then the answer to this problem is $| \overline{A_5} \cap \overline{A_7} |$. Applying the sieve formula, we get

$$
\begin{aligned}
| \overline{A_5} \cap \overline{A_7} | &= | S | - | A_5 | - | A_7 | + | A_5 \cap A_7 | \\
&= 999 - \left[\frac{999}{5} \right] - \left[\frac{999}{7} \right] + \left[\frac{999}{5 \times 7} \right] \\
&= 999 - 199 - 142 + 28 = 686.
\end{aligned}
$$

Example 10   (Bernoulli-Euler problem of misaddressed letters) How many ways to distribute n distinct letters into n corresponding envelops so that no letter gets to its corresponding envelops?

Solution Removing the words "letter" and "envelops" from this problem, we really want to know that how many permutations of $\{1, 2, \ldots, n\}$ there are such that k is not at k^{th} place for any k $(1 \leqslant k \leqslant n)$? These permutations are called the derangements, and we denote the number of derangements by D_n.

Let S be the set of permutations of $\{1, 2, \ldots, n\}$ and A_i the set of permutations $\{a_1, a_2, \ldots, a_n\}$ of $\{1, 2, \ldots, n\}$ satisfying $a_i = i$ $(i = 1, 2, \ldots, n)$. Obviously, we have

$$|S| = n!, \ |A_i| = (n-1)!, \ |A_i \cap A_j| = (n-2)!, \cdots,$$

$$|A_{i_1} \cap A_{i_2} \cap \cdots \cap A_{i_k}| = (n-k)! \, (1 \leqslant i_1 < i_2 < \cdots < i_k \leqslant n).$$

By the sieve formula, we get

$$D_n = |\overline{A_1} \cap \overline{A_2} \cap \cdots \cap \overline{A_n}| = |S| - |A_1 \cup A_2 \cup \cdots \cup A_n|$$

$$= |S| - \sum_{i=1}^{n} |A_i| + \sum_{1 \leqslant i < j \leqslant n} |A_i \cap A_j|$$

$$- \sum_{1 \leqslant i < j < k \leqslant n} |A_i \cap A_j \cap A_k| \cdots + (-1)^n |A_1 \cap A_2 \cap \cdots \cap A_n|$$

$$= n! - \binom{n}{1}(n-1)! + \binom{n}{2}(n-2)! - \binom{n}{3}(n-3)!$$

$$+ \cdots + (-1)^n \binom{n}{n} 0!$$

$$= n! \left(1 - \frac{1}{1!} + \frac{1}{2!} - \frac{1}{3!} + \cdots + \frac{(-1)^n}{n!}\right).$$

Permutation and its fixed point Let $X = \{1, 2, \ldots, n\}$, φ be a bijective mapping between X and X. Then φ is called a permutation on X and we usually write a permutation as follows:

$$\begin{pmatrix} 1 & 2 & 3 & \cdots & n \\ \varphi(1) & \varphi(2) & \varphi(3) & \cdots \varphi(n) \end{pmatrix}.$$

For $i \in X$, if $\varphi(i) = i$, then i is called a fixed point of permutation φ on X.

From example 10, we have the following corollary.

Corollary　The number of permutations with no fixed point of X is equal to

$$D_n = n!\left(1 - \frac{1}{1!} + \frac{1}{2!} - \frac{1}{3!} + \cdots + \frac{(-1)^n}{n!}\right).$$

Example 11　Suppose that $X = \{1, 2, \ldots, n\}$ and denote the number of permutations with no fixed point of X by f_n, the number of permutations with exactly one fixed point of X by g_n. Prove $|f_n - g_n| = 1$. (The 14th Canadan Mathematical Olympiad)

Proof　Let g_{ni} denote the number of permutations with exactly one fixed point i $(i = 1, 2, \ldots, n)$, then

$$g_n = g_{n1} + g_{n2} + \cdots + g_{nn}.$$

By the above corollary, we have

$$f_n = D_n, \; g_{ni} = D_{n-1}(i = 1, 2, \ldots, n) \text{ and } g_n = nD_{n-1}.$$

Hence

$$
\begin{aligned}
|f_n - g_n| &= |D_n - nD_{n-1}| \\
&= \left| n!\left(1 - \frac{1}{1!} + \frac{1}{2!} - \frac{1}{3!} + \cdots + \frac{(-1)^n}{n!}\right) \right. \\
&\quad \left. - n \cdot (n-1)!\left(1 - \frac{1}{1!} + \frac{1}{2!} - \frac{1}{3!} + \cdots + \frac{(-1)^{n-1}}{(n-1)!}\right) \right| \\
&= \left| n! \cdot \frac{(-1)^n}{n!} \right| = 1.
\end{aligned}
$$

Example 12　A new sequence $\{a_n\}$ is obtained from the sequence of the positive integers $\{1, 2, 3, \ldots\}$ by deleting all multiples of 3 or 4 except 5. Evaluate a_{2009}.

Solution Ⅰ (Estimate Value Method)　Let $a_{2009} = n$, $S = \{1, 2, \ldots, n\}$, and $A_i = \{k \mid k \in S, k \text{ is divisible by } i\}$ $(i = 3, 4, 5)$, then the set of numbers which are not deleted is $(\overline{A_3} \cap \overline{A_4} \cap \overline{A_5}) \cup A_5$. Applying the sieve formula, we get

$$
\begin{aligned}
2009 &= |(\overline{A_3} \cap \overline{A_4} \cap \overline{A_5}) \cup A_5| \\
&= |\overline{A_3} \cap \overline{A_4} \cap \overline{A_5}| + |A_5|
\end{aligned}
$$

$$= |S| - |A_3| - |A_4| - |A_5| + |A_3 \cap A_4| + |A_3 \cap A_5|$$
$$+ |A_4 \cap A_5| - |A_3 \cap A_4 \cap A_5| + |A_5| \qquad \text{①}$$
$$= n - \left[\frac{n}{3}\right] - \left[\frac{n}{4}\right] + \left[\frac{n}{3 \times 4}\right] + \left[\frac{n}{3 \times 5}\right] + \left[\frac{n}{4 \times 5}\right] - \left[\frac{n}{3 \times 4 \times 5}\right].$$

Applying the inequality $\alpha - 1 < [\alpha] \leqslant \alpha$, we obtain

$$2009 < n - \left(\frac{n}{3} - 1\right) - \left(\frac{n}{4} - 1\right) + \frac{n}{3 \times 4}$$
$$+ \frac{n}{3 \times 5} + \frac{n}{4 \times 5} - \left(\frac{n}{3 \times 4 \times 5} - 1\right) \qquad \text{②}$$
$$= \frac{3}{5}n + 3,$$

and

$$2009 > n - \frac{n}{3} - \frac{n}{4} + \left(\frac{n}{3 \times 5} - 1\right) + \left(\frac{n}{3 \times 5} - 1\right)$$
$$+ \left(\frac{n}{4 \times 5} - 1\right) - \frac{n}{3 \times 4 \times 5} \qquad \text{③}$$
$$= \frac{3}{5}n - 3.$$

Uniting ② and ③, we get $3343 \frac{1}{3} < n < 3353 \frac{1}{3}$.

If n is the multiple of 3 or 4 but not 5, then n is not a term in new sequence $\{a_n\}$, so the required n is only one of the following numbers:

$$3345,\ 3346,\ 3347,\ 3349,\ 3350,\ 3353.$$

Substituting these numbers to the equation ①, we know that $n = 3347$ is the solution of equation ①, and the answer to this problem is unique. Hence $a_{2009} = 3347$.

Solution Ⅱ (Combinatorial Analysis Method) Since the least common multiple of 3, 4 and 5 is 60. Let $S_0 = \{1, 2, \ldots, 60\}$, $A_i = \{k \mid k \in S_0,\ k \text{ is divisible by } i\}$, $(i = 3, 4, 5)$, then the set of numbers which are not deleted in S_0 is $(\overline{A_3} \cap \overline{A_4} \cap \overline{A_5}) \cup A_5$. Applying the sieve formula, we get

$$| (\overline{A_3} \cap \overline{A_4} \cap \overline{A_5}) \cup A_5 | = | \overline{A_3} \cap \overline{A_4} \cap \overline{A_5} | + | A_5 |.$$
$$= | S | - | A_3 | - | A_4 | - | A_5 | + | A_3 \cap A_4 |$$
$$+ | A_3 \cap A_5 | + | A_4 \cap A_5 |$$
$$- | A_3 \cap A_4 \cap A_5 | + | A_5 |$$
$$= 60 - \left[\frac{60}{3} \right] - \left[\frac{60}{4} \right] + \left[\frac{60}{3 \times 4} \right] + \left[\frac{60}{3 \times 5} \right]$$
$$+ \left[\frac{60}{4 \times 5} \right] - \left[\frac{60}{3 \times 4 \times 5} \right]$$
$$= 36.$$

Hence there are 36 terms of new sequence $\{a_n\}$ in S_0:

$$a_1 = 1, \ a_2 = 2, \ a_3 = 5, \ a_4 = 7, \ \cdots, \ a_{36} = 60.$$

Let $P = \{a_1, a_2, \ldots, a_{36}\}$ and $a_n = 60k + r(k, r$ are the nonnegative integers and $1 \leqslant r \leqslant 60)$. Since $(a_n, 12) = (60k + r, 12) = (r, 12) = 1$, or $(a_n, 12) = (r, 12) \neq 1$, but $5 | a_n$, then $5 | r$. Hence $r \in P$.

On the other hand, for any positive integer with the form as $60k + r$ $(k, r$ are the nonnegative integers and $r \in P)$. If $(r, 12) = 1$, so $(60k + r, 12) = 1$, thus $60k + r$ is a term of new sequence $\{a_n\}$. If $(r, 12) \neq 1$, then $5 | r$ (since $r \in P$), so $5 | 60k + r$, then $60k + r$ is also a term of new sequence $\{a_n\}$.

Therefore new sequence $\{a_n\}$ consist of all positive numbers with the form as $60k + r$ $(k, r$ are the nonnegative integers and $r \in P)$. For the given k, we obtain 36 successive terms of new sequence $\{a_n\}$ as r ranges over the set P. Note $2009 = 36 \times 55 + 29$, so $a_{2009} = 60 \times 55 + a_{29}$. But

$$a_{36} = 60, \ a_{35} = 59, \ a_{34} = 58, \ a_{33} = 55,$$
$$a_{32} = 53, \ a_{31} = 50, \ a_{30} = 49, \ a_{29} = 47,$$

thus $a_{2009} = 3300 + 47 = 3347$.

Exercise 1

1 A teacher gave out $n + 1$ prizes to n students such that each student has at least one prize. Then the number of distinct sending ways is (　).

(A) $n\mathrm{P}_n^{n+1}$ (B) $(n+1)\mathrm{P}_n^n$ (C) P_n^{n+1} (D) $\binom{n+1}{2}\mathrm{P}_n^n$

2 Suppose that a teacher selects 4 students from 5 boys and 4 girls to form a debate team. If at least one boy and one girl must be selected, then the number of distinct selecting ways is (　).

(A) 60 (B) 80 (C) 120 (D) 420

3 If the 5-digit numbers greater than 20000 which are not the multiples of 5 have the following properties: their digits are distinct and each digit is one of the numbers 1, 2, 3, 4, 5, then the number of these 5-digit numbers is (　).

(A) 96 (B) 76 (C) 72 (D) 36

4 If the coefficients A and B of the equation of a straight line $Ax+By=0$ are two distinct digits from the numbers 0, 1, 2, 3, 6, 7, then the number of distinct straight lines is _____.

5 If the base a and the variable x of the logarithm $\log_a x$ are two distinct digits from 1, 2, 3, 4, 5, 7, 9, then the number of distinct values of the logarithm $\log_a x$ is _____.

6 In a table tennis tournament, each player plays exactly one game against each of the other players. But during this process, there are 3 players who have withdrawn from the tournament and each of them participates in exactly two matches. If the total of matches is 50, then the number of matches whin the above 3 players is (　).

(A) 0 (B) 1 (C) 2 (D) 3

(China Mathematical Competition in 1994)

7 Suppose that a, b, c in the equation of straight line $ax+by+c=0$ are three distinct elements of set $\{-3, -2, -1, 0, 1, 2, 3\}$ and the inclination of straight line is an acute angle. Then the number of distinct straight lines is _____. (China Mathematical Competition in 1999)

8 A 2×3 rectangle is divided into six unit squares A, B, C, D, E, F. Each of these unit squares is to be colored in one of 6 colors such that no two adjacent squares have the same colors. Then the number of distinct coloring ways is _____.

9 Two teams A and B participate in a table tennis tournament. There are 7 players of each team to engage in this tournament in a determined order. Firstly, 1st player of A team plays against 1st player of B team and the loser is eliminated. Afterward, the winner plays against 2nd player of another team. On subsequent steps, the play is similar. Thus the game does not end until all players of some team are eliminated, and another team wins. Then the number of the distinct processes of game is _____. (China Mathematical Competition in 1988)

10 In a shooting tournament, eight clay targets are arranged in two hanging columns of three each and one column of two, as pictured. A marksman is to break all eight targets according to the following rules: (1) The marksman first chooses a column from which a target is to be broken. (2) The marksman

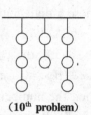

(10th **problem**)

must then break the lowest remaining unbroken target in the chosen column. If these ruses are flowed, in how many different orders can the eight targets be broken. (8th American Invitational Mathematical Examination in 1990)

11 How many ways are there to paint the five vertices of a regular quadrangular pyramid with 5 colors such that each vertex is exactly painted with one of 5 colors and the vertices with a common edge must be painted with different colors?

(**Remark** A coloring is the same as another which is from the rotation of the former).

12 It is given that there are two sets of real numbers $A = \{a_1, a_2, \ldots, a_{100}\}$ and $B = \{b_1, b_2, \ldots, b_{50}\}$. If there is a mapping f from A to B such that every element in B has an inverse image and $f(a_1) \leqslant f(a_2) \leqslant \cdots \leqslant f(a_{100})$, then the number of such mappings is (   ).

(A) $\binom{100}{50}$ (B) $\binom{98}{50}$ (C) $\binom{100}{49}$ (D) $\binom{99}{49}$

（China Mathematical Competition in 2002）

13 A natural number a is called a "lucky number" if the sum of its digits is 7. Arrange all "lucky numbers" in an ascending order, and we get a sequence a_1, a_2, a_3,.... If $a_n = 2005$, then $a_{5n} = $ _____ .
（China Mathematical Competition in 2005）

14 How many ways are there to arrange n married couples in a line such that no man is adjacent to his wife?

15 Suppose that all positive integers which are relatively prime to 105 are arranged into a increasing sequence: a_1, a_2, a_3,.... Evaluate a_{1000}. （China Mathematical Competition in 1994）

16 How many n-digit numbers are there consisting of the digits 1, 2, 3 with at least one 1, at least one 2 and at least one 3?

Chapter 2

Pigeonhole Principle and Mean Value Principle

2. 1   Pigeonhole Principle

Pigeonhole Principle also is called Drawer Principle or Dirichlet's Principle. Pigeonhole Principle is a basic and important principle of combinatorial mathematics which is widely used to prove many existence problems.

The First Pigeonhole Principle   If m objects are put into n boxes, then at least one box contains $\left[\dfrac{m-1}{n}\right]+1$ or more objects, where $[x]$ denotes the largest integer less than or equal to x.

Proof   The proof is by contradiction. If each of n boxes contained at most $\left[\dfrac{m-1}{n}\right]$ of the objects, then the total number of objects would be at most $n\left[\dfrac{m-1}{n}\right] \leqslant n \cdot \dfrac{m-1}{n} = m-1$. Since we started with m objects, this is impossible.

Corollary of The First Pigeonhole Principle   Let $m_1, m_2, \ldots, m_n$ be n positive integers. If $m_1 + m_2 + \cdots + m_n + 1$ objects are put into n boxes, then the first box contains at least $m_1 + 1$ objects, or the second box contains at least $m_2 + 1$ objects, $\ldots$, or the nth box contains at least $m_n + 1$ objects.

Proof   Proof by contradiction. If the first box contains at most m_1 objects, and the second box contains at most m_2 objects, $\ldots$, and the nth box contains at most m_n objects, then the total of objects would be at most $m_1 + m_2 + \cdots + m_n$ Since we start with $m_1 + m_2 + \cdots +$

$m_n + 1$ objects, it is impossible.

The Second Pigeonhole Principle   If m objects are put into n boxes, then at least one box contains $\left[\dfrac{m}{n}\right]$ or less objects.

Proof   Proof by contradiction. If each of n boxes contains at least $\left[\dfrac{m}{n}\right]+1$ objects, then the total of objects would be least $n\left(\left[\dfrac{m}{n}\right]+1\right) >$ $n \cdot \dfrac{m}{n} = m$. Since we start with m objects, it is impossible.

Corollary of The Second Pigeonhole Principle   Let m_1, m_2, ..., m_n be n positive integers. If $m_1 + m_2 + \cdots + m_n - 1$ objects are put into n boxes, then the first box contains at most $m_1 - 1$ objects, or the second box contains at most $m_2 - 1$ objects,..., or the nth box contains at most $m_n - 1$ objects.

Proof   Proof by contradiction. If the first box contains at least m_1 objects, and the second box contains at least m_2 objects,..., and the nth box contains at least m_n objects, then the total of objects would be at least $m_1 + m_2 + \cdots + m_n$ Since we start with $m_1 + m_2 + \cdots + m_n - 1$ objects, it is impossible.

Example 1 (Ramsey's Theorem)   There are six points in a 3-dimensional space no four points of which are coplanar. Two distinct points determine a line segment, and let each line segment be colored red or blue. Then there exist a triangle such that the 3 sides of this triangle are all colored red or blue.

Proof   Consider a given point A. By Pigeonhole Principle, of the 5 line segments meeting A, either at least $\left[\dfrac{5-1}{2}\right]+1 = 3$ are colored red or at least 3 are colored blue. For the sake of argument let us suppose that there are 3 line segments AB, AC and AD are colored red. Consider the triangle BCD, if a side of $\triangle BCD$, say BD, is colored red, then the 3 sides of $\triangle ABD$ are all colored red. Otherwise, the 3 sides of $\triangle BCD$ all are colored blue. This completes the proof.

It seems convenient to list some basic concepts of the graph theory that will be used throughout the book. The set V_n of n points in a plane is called the set of vertices. A complete graph on n vertices, denoted by K_n, consists of a set V_n of vertices and a set E of edges (or sides) connecting each pair of distinct vertices in V. A closed broken line which consists of m edges of K_n is called an m-sided polygon, and the 3-sided polygon is also called a triangle. Let each of edges of a complete graph K_n be colored in one of m colors, then this graph K_n is called an m-colored complete graph. Let each side of an m-sided polygon P be colored red (or blue), then this m-sided polygon P is called a red (or blue) m-sided polygon. Applying the concepts of the graph theory, the Ramsey's theorem can be described equivalently as follows.

Ramsey's Theorem   There exists a monochromatic triangle in a 2-colored complete graph K_6.

On the other hand, as in figure 2.1, we can exhibit a 2-colored complete graph K_5 which has no a red triangle no a blue triangle (the solid lines colored red and the dashed lines colored blue). Generally, if there exists a smallest positive integer n with the following property: there exists a monochromatic

Figure 2.1

triangle in an m-colored complete graph K_n, then the smallest positive integer n is called Ramsey's number, denoted by R_m. So we get $R_2 = 6$ from the above conclusion. But it is very difficult to find the Ramsey's numbers. We only know a few Ramsey's numbers: $R_2 = 6$, $R_3 = 17$, $R_4 = 65$, ...

Suppose that we represent a people with a point. If two people are mutually acquainted, then two corresponding points are connected by a red line segment, and if two people are mutually unacquainted, then two correspondent points are connected by a blue line segment. Thus the above Ramsey's theorem becomes the following Hangary's mathematical competition problem: Prove that among any 6 persons there exist 3 persons who form $\binom{3}{2} = 3$ pair of mutually acquainted or mutually unacquainted persons. (Acquaintance is a symmetric relation.)

Example 2   Let k be a given positive integer number. Find the smallest value of n such that among any n positive integer numbers there exist two numbers whose sum or difference is divisible by $2k$.

Solution   Let a_1, a_2, ..., a_n be any n positive integers and their remainders modulo $2k$ be r_1, r_2, ..., r_n, then $r_i \in S = \{0, 1, 2, ..., 2k - 1\}$. Consider the following classifications of S:

$$. \{0\}, \{1, 2k - 1\}, \{2, 2k - 2\}, ..., \{k - 1, k + 1\}, \{k\}.$$

If $n \geqslant k + 2$, by the pigeonhole principle, among n integers r_1, r_2, ..., r_n there exists two numbers r_i, r_j $(i \neq j, i, j \in S)$ which belong to the same group. So the sum or difference of a_i and a_j is divisible by $2k$.

On the other hand, it is not difficult to show that among $k + 1$ numbers $0, 1, 2, ..., k$ there are not two numbers a and b such that $a + b$ or $a - b$ is divisible by $2k$. It follows that the required smallest value is $k + 2$.

Example 3   Let α be a positive real number and n be a positive integer. Prove that there exist two positive integers p and q satisfying the following inequality: $\left| \alpha - \dfrac{q}{p} \right| \leqslant \dfrac{1}{np}$.

Proof   Firstly, we prove the following proposition: Given $n + 1$ real numbers x_0, x_1, ..., $x_n \in [0, 1)$, there exist i, j $(0 \leqslant i < j \leqslant n)$ such that $| x_i - x_j | < \dfrac{1}{n}$. In fact, suppose that the interval $[0, 1)$ is divided into n small intervals: $\left[0, \dfrac{1}{n}\right), \left[\dfrac{1}{n}, \dfrac{2}{n}\right), ..., \left[\dfrac{n-1}{n}, 1\right)$. By Pigeonhole Principle, there exist x_i, x_j $(0 \leqslant i < j \leqslant n)$ which belong to same small interval. Thus $| x_i - x_j | < \dfrac{1}{n}$.

Let $m_i = [i\alpha]$, $i = 0, 1, ..., n$, then $m_i \leqslant i\alpha < m_i + 1$, i.e. $0 \leqslant i\alpha - m_i < 1$. By the above proposition, there exist $0 \leqslant k < l \leqslant n$ such that $| (l\alpha - m_l) - (k\alpha - m_k) | < \dfrac{1}{n}$, i.e. $| (l - k)\alpha - (m_l - m_k) | <$

$\dfrac{1}{n}$. Set $p = l - k$, $q = m_l - m_k$, then p, q are positive integers and

$$\left| \alpha - \frac{q}{p} \right| \leqslant \frac{1}{np}.$$

Remark Since $p \geqslant 1$, $\left| \alpha - \dfrac{q}{p} \right| \leqslant \dfrac{1}{n}$. In other words, any real number can be approximated to arbitrary precision by a ration number.

Example 4 Suppose that n numbers are deleted in the set $S = \{1, 2, \ldots, 2005\}$ such that among the remaining numbers, no number equals to the product of other two. Find the smallest positive integers n with the above properties.

Solution Consider the following 43 triples of S: $\{44, 45, 44 \times 45\}$, $\{43, 46, 43 \times 46\}$, $\{42, 47, 42 \times 47\}$, $\ldots$, $\{3, 86, 3 \times 86\}$, $\{2, 87, 2 \times 87\}$. If the number of positive integers which are deleted in the S is less than 43, then among the remaining numbers, there exist three numbers in a triple such that one of them equals the product of the other two. Hence, we must delete at least 43 elements in S.

On the other hand, if we delete the following 43 number: 2, 3, 4, $\ldots$, 44 in S. Since the product of any two of the remaining numbers is at least $45 \times 46 = 2070$ or $1 \times a = a$ ($a \in \{45, 46, \ldots, 2005\}$), among the remaining numbers, no number equals the product of the other two. Therefore the required smallest positive integers n is 43.

Example 5 Suppose that $a_1, a_2, \ldots, a_n$ ($n \geqslant 4$) are n distinct integers in open interval $(0, 2n)$. Prove that there exists a non-empty subset S of $\{a_1, a_2, \ldots, a_n\}$ such that the sum of all elements in S is divisible by $2n$.

Proof (1) Case 1. $n \notin \{a_1, a_2, \ldots, a_n\}$. Since $2n$ integers $a_1, a_2, \ldots, a_n, 2n - a_1, 2n - a_2, \ldots, 2n - a_n \in \{1, 2, \ldots, 2n - 1\}$, by Pigeonhole Principle, it follows that there exist two of them who are equal. But no two of the numbers $a_1, a_2, \ldots, a_n$ are equal and no two of the numbers $2n - a_1, 2n - a_2, \ldots, 2n - a_n$ are equal, so there is an

i and a j such that $a_i = 2n - a_j$. Since $a_i \neq n$, $a_j \neq n$, $i \neq j$ and $a_i + a_j = 2n$. In this case, the conclusion is valid.

(2) Case 2. $n \in \{a_1, a_2, \ldots, a_n\}$. Without loss of generality, let $a_n = n$. Consider $n - 1$ numbers $a_1, a_2, \ldots, a_{n-1} (n - 1 \geqslant 3)$. If for any 3 numbers $a_i < a_j < a_k$, $a_j - a_i$ and $a_k - a_j$ are divisible by n, then $a_k - a_i = (a_k - a_j) + (a_j - a_i) \geqslant 2n$. It is impossible. Hence there exist two numbers $a_i < a_j (i, j \in \{1, 2, \ldots, n - 1\})$ such that $a_j - a_i$ is not divisible by n. Without loss of generality, let $a_2 - a_1$ be not divisible by n. Consider the following n numbers:

$$a_1, a_2, a_1 + a_2, a_1 + a_2 + a_3, \ldots, a_1 + a_2 + \cdots + a_{n-1}.$$

(Ⅰ) If the n numbers module n are distinct, then there exists one of these numbers which is divisible by n. Suppose that this number is kn ($k \in \mathbf{N}_+$). If k is an even, then the conclusion is valid. If k is an odd, then $k + 1$ is an even and $kn + a_n = (k + 1)n$ is divisible by $2n$. The conclusion is also valid.

(Ⅱ) If there are two of the n numbers which are congruent with module n, then their difference is divisible by n. But $a_2 - a_1$ is not divisible by n. Hence this difference is equal to the sum of some (at least one) of the numbers $a_1, a_2, \ldots, a_{n-1} (n - 1 \geqslant 3)$. Thus this conclusion is also valid as (Ⅰ).

Remark In this problem, the condition $n \geqslant 4$ is necessary. When $n = 3$, for set $\{1, 3, 4\}$ the conclusion of this problem is not valid.

Example 6 Let 49 students solve 3 problems and the score of each problem is one of these nonnegative integers 0, 1, 2, 3, 4, 5, 6, 7. Prove that there exist two students A and B such that for each of the problems the scores of A is not less than B. (The Problem Prepared for 29th IMO in 1988)

Proof Suppose that there are two students A and B such that for first and second problem, their scores are the same. If the score of the third problem A get is not less than B's, then for each of the problems the scores of A is not less than B's.

Now, suppose that for any two students, their scores of first and second problems are not all the same and we represent each of the 49 students by a point (i, j) $(0 \leqslant i, j \leqslant 7)$ in a plane where i, j are his (or her) scores of the first and second problems respectively. Thus the 49 students correspond to 49 distinct integer points in plane. Set

$$S = \{(i, j) \mid i, j \text{ are integer numbers and } 0 \leqslant i, j \leqslant 7\};$$

$$M_1 = \{(i, j) \mid (i, j) \in S, 0 \leqslant i \leqslant 7, j = 0 \text{ or } i = 7, 1 \leqslant j \leqslant 7\};$$

$$M_2 = \{(i, j) \mid (i, j) \in S, 0 \leqslant i \leqslant 6, j = 1 \text{ or } i = 6, 2 \leqslant j \leqslant 7\};$$

$$M_3 = \{(i, j) \mid (i, j) \in S, 0 \leqslant i \leqslant 5, j = 2 \text{ or } i = 5, 3 \leqslant j \leqslant 7\};$$

$$M_4 = \{(i, j) \mid (i, j) \in S, 0 \leqslant i \leqslant 4, j = 3 \text{ or } i = 4, 4 \leqslant j \leqslant 7\};$$

$$M_5 = \{(i, j) \mid (i, j) \in S, i = 2, 3 \text{ and } 4 \leqslant j \leqslant 7\};$$

$$M_6 = \{(i, j) \mid (i, j) \in S, i = 0, 1 \text{ and } 4 \leqslant j \leqslant 7\}.$$

Thus the 49 integer points belong to the set

$$S = M_1 \cup M_2 \cup M_3 \cup M_4 \cup M_5 \cup M_6.$$

By Pigeonhole Principle, there are at least $\left[\dfrac{49-1}{6}\right] + 1 = 9$ integral points all of which belong to the same set, denoted by M. Since $|M_5| = |M_6| = 8$, where $|M_i|$ denote the numbers of elements in M_i, M is one of M_1, M_2, M_3 or M_4. Since the 9 integral points in M correspond 9 distinct students whose score of the third problem are 0, 1, 2, 3, 4, 5, 6, 7, there are at least two students whose scores of the third problem are the same. By the constructions of M_1, M_2, M_3 or M_4, we know that there exists one (denoted by A) of the two students, such that for the first and second problems, the scores of A are not less than another student's (denoted by B). Therefore, for each problem, the scores of A are not less than B's.

Example 7 Let n and r be two positive integers. Find the smallest positive integer m satisfying the following condition: For each classification of the set $\{1, 2, \ldots, m\}$ into r subsets $A_1, A_2, \ldots, A_r$ $(A_i \cap A_j = \varnothing, i \neq j)$, there exist two numbers a and b in a subset

$A_i (1 \leqslant i \leqslant r)$ such that $1 \leqslant \dfrac{b}{a} \leqslant 1 + \dfrac{1}{n}$.

Proof Denote the smallest value of m by m_0 (the value of m_0 can be obtained from the following analysis). Suppose that $m < m_0$. Let $A_j = \{ i \mid i \equiv j \pmod{r}, i \leqslant m \} (j = 1, 2, \ldots, r)$. Then for any two numbers a, b in a subset with $a > b$, we have $b \leqslant a - r < m_0 - r$. Thus

$$\frac{a}{b} = 1 + \frac{a-b}{b} > 1 + \frac{r}{m_0 - r}.$$

Hence when $m_0 = nr + r$, we have $\dfrac{a}{b} > 1 + \dfrac{1}{n}$. Therefore the condition in the question cannot hold. Next, For $m = m_0 = nr + r$, consider the $(r+1)$ numbers $nr, nr + 1, \ldots, nr + r$. By Pigeonhole Principle, there exists $a > b$ such that a and b are in the same subset. Then $a - b \leqslant r$, $b \geqslant nr$, so

$$1 < \frac{a}{b} = 1 + \frac{a-b}{b} \leqslant 1 + \frac{r}{b} \leqslant 1 + \frac{r}{nr} = 1 + \frac{1}{n}.$$

Therefore, the required smallest value of m is $m_0 = nr + r$.

Example 8 Let a space figure G consist of $n (\geqslant 4)$ vertices, no 4 vertices of which are coplanar and there are $\left[\dfrac{n^2}{4} \right] + 1$ line segments connecting these vertices. Prove there exist two triangles with a common side in the figure G.

Proof We will prove the statement by induction on n. For the basic case $n = 4$, suppose the 4 vertices are A, B, C, D and there are $\left[\dfrac{4^2}{4} \right] + 1 = 5$ line segments connecting these vertices. Thus only $\dbinom{4}{2} - 5 = 1$ pair of these vertices is not connected by a line segment. Without loss of generality, suppose that there is not a line segment connecting the vertices C and D. In this case, there exist two triangles ABC and ABD with a common side AB. Thus the basic case is proved.

Now we assume that the statement is true for $n = k \geqslant 4$. For $n = k + 1$, let a space figure G consist of $k + 1$ vertices and there are

$\left[\frac{(k+1)^2}{4}\right]+1$ line segments connecting these vertices. Then the sum

of the line segments meeting each vertex equals $2\left(\left[\frac{(k+1)^2}{4}\right]+1\right)$. By

Pigeonhole Principle, there are at most $\left[\frac{2}{k+1}\left(\left[\frac{(k+1)^2}{4}\right]+1\right)\right]$ line

segments meeting some vertex A. By deleting the vertex A and all line
segments meeting vertex A, then among residual figure, there are k
vertices and at least

$$N = \left[\frac{(k+1)^2}{4}\right]+1-\left[\frac{2}{k+1}\left(\left[\frac{(k+1)^2}{4}\right]+1\right)\right]$$

line segments connecting these vertices.

If $k = 2m \ (m \geqslant 2)$ is even, then

$$N = m(m+1)+1-\left[m+\frac{m+2}{2m+1}\right]$$
$$= m(m+1)+1-m$$
$$= m^2+1 = \left[\frac{k^2}{4}\right]+1.$$

If $k = 2m-1 \ (m \geqslant 3)$ is odd, then

$$N = m^2+1-\left[m+\frac{1}{m}\right] = m(m-1)+1 = \left[\frac{k^2}{4}\right]+1,$$

In a word, there are at least $\left[\frac{k^2}{4}\right]+1$ line segments connecting k

vertices. By the inductive hypothesis, there exist two triangles with a
common side. Thus the statement is proved.

    Remark   If $n(\geqslant 4)$ is even, then this becomes a question in the 2nd
selection examination for the national team of China.

2. 2   The Mean Value Principle

    The mean value principle   (1) Let $a_1, a_2, \ldots, a_n$ be n real

numbers and $A = \frac{1}{n}(a_1 + a_2 + \cdots + a_n)$, then at least one of the numbers a_1, a_2, ... , a_n is greater than or equal to A and also at least one of these numbers is less than or equal to A.

(2) Let a_1, a_2, ... , a_n be n positive real numbers and $G = \sqrt[n]{a_1 a_2 \ldots a_n}$, then at least one of the numbers a_1, a_2, ... , a_n is gerater than or equal to G and also at least one of these numbers is less than or equal to G.

Proof (1) $\underset{1 \le i \le n}{\text{Min}}\{a_i\} \le A \le \underset{1 \le i \le n}{\text{Max}}\{a_i\}$, (2) $\underset{1 \le i \le n}{\text{Min}}\{a_i\} \le G \le \underset{1 \le i \le n}{\text{Max}}\{a_i\}$.

Example 9 Suppose that 10 distinct numbers 1, 2, 3, 4, 5, 6, 7, 8, 9, 10 are arranged in a circle with any order. Show that there exist 3 successive numbers whose sum is not less than 18.

Proof Suppose that a_1, a_2, ... , a_{10} are 10 numbers 1, 2, 3, 4, 5, 6, 7, 8, 9, 10 in a circle. Without loss of generality, let $a_1 = 1$, then $\frac{1}{3}[(a_2 + a_3 + a_4) + (a_5 + a_6 + a_7) + (a_8 + a_9 + a_{10})] = \frac{1}{3}(2 + 3 + 4 + \cdots + 10) = \frac{54}{3} = 18$. By the mean value principle, one of $a_2 + a_3 + a_4$, $a_5 + a_6 + a_7$ and $a_8 + a_9 + a_{10}$ is not less than 18.

Example 10 Suppose that there are n (≥ 4) distinct points in the plane and each pair of them is connected by a line segment. If among all lengths of these line segments, only $n + 1$ lengths equal d. Prove that there exists a point P such that there are at least 3 line segments meeting P their lengths all equal d.

Proof Let n points be P_1, P_2, ... , P_n and the number of the line segments meeting P_i with length d be $d_i (i = 1, 2, \ldots, n)$. Then

$$d_1 + d_2 + \cdots + d_n = 2(n + 1)$$

and

$$\frac{1}{n}(d_1 + d_2 + \cdots + d_n) = \frac{2(n + 1)}{n} > 2.$$

By the mean value principle, there exists $d_i \ge 3$, i.e. there are at least 3 line segments meeting P_i their lengths all equal d.

Example 11 Let $f(z) = c_0 z^n + c_1 z^{n-1} + \cdots + c_{n-1} z + c_n$ be a polynomial of n degree with complex coefficients. Prove that there exists a complex number z_0 such that $|z_0| \leqslant 1$ and $|f(z_0)| \geqslant |c_0| + |c_n|$.

Proof I Let $\omega = \cos\dfrac{2\pi}{n} + i\sin\dfrac{2\pi}{n}$, $\omega_k = \omega^k = \cos\dfrac{2k\pi}{n} + i\sin\dfrac{2k\pi}{n}$ $(k = 0, 1, \ldots, n-1)$, $\alpha = \cos\theta + i\sin\theta$. (We can determine the value of θ from the following analysis.) Then $\omega^0 = \omega^n = 1$, $\omega^j \neq 1$ $(j = 2, 3, \ldots, n-1)$ and

$$\sum_{k=0}^{n-1} \omega_k^j = n \ (j = 0 \text{ or } n),$$

$$\sum_{k=0}^{n-1} \omega_k^j = \sum_{k=0}^{n-1} \omega^{jk} = \frac{1 - (\omega_j)^n}{1 - \omega_j} = 0 \ (j = 2, 3, \ldots, n-1),$$

$$\sum_{k=0}^{n-1} f(\alpha\omega_k) = c_0\alpha^n \sum_{k=0}^{n-1} \omega_k^n + c_1\alpha^{n-1} \sum_{k=0}^{n-1} \omega_k^{n-1} + \cdots + c_{n-1}\alpha \sum_{k=0}^{n-1} \omega_k + c_n n$$

$$= n(c_0\alpha^n + c_n).$$

We may select the argument θ of α such that the principle values of the arguments of $c_0\alpha^n$ and c_n are equal. (In fact, let the principle values of the arguments of c_0 and c_n be θ_0 and θ_n respectively, then we set $\theta = \left(\dfrac{1}{n}(\theta_n - \theta_0)\right)$.) Hence

$$\sum_{k=0}^{n-1} |f(\alpha\omega_k)| \geqslant \left|\sum_{k=0}^{n-1} f(\alpha\omega_k)\right|$$

$$= n|c_0\alpha^n + c_n|$$

$$= n(|c_0| \cdot |\alpha|^n + |c_n|)$$

$$= n(|c_0| + |c_n|).$$

By the mean value principle, there exists k_0 such that

$$|f(\alpha\omega_{k_0})| \geqslant \frac{1}{n} \sum_{k=0}^{n-1} |f(\alpha\omega_k)| = |c_0| + |c_n|.$$

Set $z_0 = \alpha\omega_{k_0}$, then $|z_0| = 1$, and $|f(z_0)| \geqslant |c_0| + |c_n|$. This completes the proof.

Proof II Let the complex number u satisfy that $|u| = |c_0| + |c_n|$

and u has the same argument with c_n. Let

$$g(z) = f(z) - u = c_0 z^n + c_1 z^{n-1} + \cdots + c_{n-1} z + c_n - u$$

and the n complex roots of $g(z)$ be $z_1, z_2, \ldots, z_n$. Thus

$$| z_1 z_2 \cdots z_n | = \frac{| c_n - u |}{| c_0 |} = \frac{|| c_n | - | u ||}{| c_0 |} = \frac{| c_0 |}{| c_0 |} = 1.$$

By the mean value principle, there are a k_0 $(1 \leqslant k_0 \leqslant n)$ such that $| z_{k_0} | \leqslant \sqrt[n]{| z_1 z_2 \ldots z_n |} = 1$. Set $z_0 = z_{k_0}$, then $| z_0 | \leqslant 1$ and $g(z_0) = f(z_0) - u = 0$, i.e. $| f(z_0) | = | u | = | c_0 | + | c_n |$. This completes the proof.

Exercise 2

1 In a group of 17 scientists each scientist sends letters to the others. In their letters only three topics are involved and each couple of scientists makes reference to only one topic. Show that there exists a group of three scientists which send each other letters on the same topic. (6th IMO)

2 Is it possible to choose (1) 4; (2) 5 distinct positive integers such that the sum of any three numbers of them is a prime?

3 Given 7 points in $\triangle ABC$, no three of which are collinear. Prove that there exists 3 points of these points which determine a triangle with the area less than or equal to $\frac{1}{4} S_{\triangle ABC}$.

4 Let $S = \{1, 2, \ldots, 2009\}$. Find the largest positive integer n such that there is an n-elements subset A of S with the following property: the sum of any two numbers in A is not divisible by their difference.

5 Let M be the subset of $S = \{1, 2, \ldots, 2000\}$ such that the differente of any two numbers of M is neither 5 nor 8. How many elements of M at most are there?

6 Suppose that each of 7 boys has 3 brothers among the

remaining 6 boys. Prove that the 7 boys are all brothers.

7   Suppose 10 persons buy 30 books at a book-shop satisfying the following conditions:

(1) Each of them buys 3 books,

(2) Among the books which were bought by any two of them, there are at least one is same. Suppose the number of purchasers who bought some book is largest. Determine what that smallest value of the largest number is? (8th CMO)

8   Every point in a plane is colored red or blue. Prove that

(1) For any positive real number a there exists a triangle with the sides of lengths a, $\sqrt{3}a$, $2a$ such that its 3 vertices all are colored red (or blue).

(2) There exist two similar triangles such that 3 vertices of each of two triangles are colored in the same color. (The question (2) is a question of China Mathematical Competition in 1995)

9   Suppose a convex polyhedron has 6 vertices and 12 edges. Prove that each facet is a triangle.

10   Let the set S be a set with finite number of point in the plane. If the distances between any two points in S is determined, then the set S is called stable. Let M_n be a set of $n(\geqslant 4)$ points in the plane and no three points of M_n be collinear. Suppose that the number of pairs of points having the determined distances in M_n is $\frac{1}{2}n(n-3) +$ 4. Show that M_n is stable. (China Shanghai Mathematical Competition in 1999)

Chapter 3　The Generating Functions

3.1　The Concept of Generating Function

Let $f(x) = (1+x)^n$. By applying the binomial theorem, we gain

$$f(x) = \sum_{k=0}^{n} \binom{n}{k} x^k = \binom{n}{0} + \binom{n}{1} x + \binom{n}{2} x^2 + \cdots + \binom{n}{n-1} x^{n-1} + \binom{n}{n} x^n.$$

Then $f(x)$ corresponds to the sequence $\left\{ \binom{n}{k}, 0 \leqslant k \leqslant n \right\}$. Thus the function $f(x) = (1+x)^n$ is called the generating function of the sequence $\left\{ \binom{n}{k} \right\}$.

In general, for the finite sequence $a_0, a_1, a_2, \ldots, a_n$, its generating function is defined to be the following polynomial:

$$f(x) = \sum_{k=0}^{n} a_k x^k = a_0 + a_1 x + a_2 x^2 + \cdots + a_n x^n.$$

For the infinite sequence $a_0, a_1, a_2, \ldots, a_n, \ldots$, its generating function is defined to be the following formal power series:

$$f(x) = \sum_{n=0}^{\infty} a_n x^n = a_0 + a_1 x + a_2 x^2 + \cdots + a_n x^n + \cdots.$$

Let $f(x) = \sum_{n=0}^{\infty} a_n x^n$, $g(x) = \sum_{n=0}^{\infty} b_n x^n$ be two formal power series, we define

(1) $f(x) = g(x)$ if and only if $a_n = b_n (n = 0, 1, 2, \ldots)$,

(2) $f(x) \pm g(x) = \sum_{n=0}^{\infty} (a_n \pm b_n) x^n$,

(3) $cf(x) = \sum_{n=0}^{\infty} (ca_n) x^n$ (c is a constant),

(4) $f(x)g(x) = \sum_{n=0}^{\infty} c_n x^n$, where $c_n = \sum_{k=0}^{n} a_k b_{n-k}$, $n = 0$, 1, 2, $\ldots$.

3. 2 Some Important Formulas

When we solve the problems using the generating functions, besides the binomial theorem, we also need the following formulas:

Formula I   (The summation formula of the infinite geometric series with the common ratio q satisfying $-1 < q < 1$)

$$\frac{1}{1-x} = \sum_{n=0}^{\infty} x^n = 1 + x + x^2 + \cdots + x^n + \cdots \quad (-1 < x < 1).$$

Formula II Let k be a positive integer, then

$$(1-x)^{-k} = \sum_{n=0}^{\infty} \binom{k+n-1}{k-1} x^n$$

$$= 1 + \binom{k}{k-1} x + \binom{k+1}{k-1} x^2 + \cdots$$

$$+ \binom{n+k-1}{k-1} x^n + \cdots \quad (-1 < x < 1).$$

With the $(k-1)$ the derivative of the formula I at x and divided by $(k-1)!$, we get the formula II.

Example 1 Let $a_0 = -1$, $a_1 = 1$, $a_n = 2a_{n-1} + 3a_{n-2} + 3^n (n \geqslant 2)$. Find a_n.

Solution Let $f(x) = a_0 + a_1 x + a_2 x^2 + \cdots + a_n x^n + \cdots$,

Then $-2xf(x) = -2a_0 x - 2a_1 x^2 - \cdots - 2a_{n-1} x^n - \cdots$,

$-3x^2 f(x) = -3a_0 x^2 - \cdots - 3a_{n-2} x^n - \cdots$.

Adding the above three equations and applying $a_0 = -1$, $a_1 = 1$, $a_n = $

$2a_{n-1} + 3a_{n-2} + 3^n \ (n \geqslant 2)$, we have

$$(1 - 2x - 3x^2)f(x) = -1 + 3x + 3^2 x^2 + \cdots + 3^n x^n + \cdots$$

$$= -1 + \frac{3x}{1 - 3x} = \frac{6x - 1}{1 - 3x},$$

$$f(x) = \frac{6x - 1}{(1 + x)(1 - 3x)^2} = \frac{A}{1 + x} + \frac{B}{(1 - 3x)^2} + \frac{C}{1 - 3x}. \qquad ①$$

Multiplying both sides of ① by $1 + x$ and setting $x = -1$, we get

$$A = \frac{6x - 1}{(1 - 3x)^2}\Big|_{x=-1} = -\frac{7}{16}.$$

Multiplying both sides of ① by $(1 - 3x)^2$ and setting $x = \frac{1}{3}$, we obtain

$$B = \frac{6x - 1}{1 + x}\Big|_{x=\frac{1}{3}} = \frac{3}{4}.$$

Multiplying both sides of ① by x and setting $x \to +\infty$, we obtain

$$0 = \lim_{x \to +\infty} \frac{x(6x - 1)}{(1 + x)(1 - 3x)^2}$$

$$= \lim_{x \to +\infty} \left(\frac{Ax}{1 + x} + \frac{Bx}{(1 - 3x)^2} + \frac{Cx}{1 - 3x}\right)$$

$$= A - \frac{1}{3}C.$$

Hence $C = 3A = -\frac{21}{16}$,

$$f(x) = -\frac{7}{16(1 + x)} + \frac{3}{4(1 - 3x)^2} - \frac{21}{16(1 - 3x)}$$

$$= -\frac{7}{16}\sum_{n=0}^{\infty}(-1)^n x^n + \frac{3}{4}\sum_{n=0}^{\infty}\binom{n+1}{1}(3x)^n - \frac{21}{16}\sum_{n=0}^{\infty}(3x)^n$$

$$= \sum_{n=0}^{\infty}\left[\frac{(4n - 3) \cdot 3^{n+1} - 7(-1)^n}{16}\right]x^n.$$

Therefore we yield

$$a_n = \frac{1}{16}\left[(4n - 3) \cdot 3^{n+1} - 7(-1)^n\right].$$

Example 2 Show that for all positive integers n, we have

$$\sum_{i=0}^{n} \binom{2n+1}{2i}\binom{2i}{i} 2^{2n-2i+1} = \binom{4n+2}{2n+1}.$$

Proof Firstly, in the equality $(1+x)^{4n+2} = \sum_{k=0}^{4n+2} \binom{4n+2}{k} x^k$, the

coefficient of x^{2n+1} is equal to $\binom{4n+2}{2n+1}$. On the other hand, in

the equality

$$(1+x)^{4n+2} = [(1+x^2)+2x]^{2n+1} = \sum_{k=0}^{2n+1} \binom{2n+1}{k}(2x)^{2n+1-k}(1+x^2)^k$$

$$= \sum_{k=0}^{2n+1} \binom{2n+1}{k} 2^{2n+1-k} x^{2n+1-k} \left(\sum_{i=0}^{k}\binom{k}{i}x^{2i}\right),$$

the coefficient of x^{2n+1} is equal to $\sum_{i=0}^{n} \binom{2n+1}{2i} 2^{2n+1-2i}\binom{2i}{i}$. Hence

$$\sum_{i=0}^{n} \binom{2n+1}{2i}\binom{2i}{i} 2^{2n-2i+1} = \binom{4n+2}{2n+1}.$$

Example 3 Show that $\displaystyle\sum_{k=0}^{[(n-1)/2]} (-1)^k \binom{n+1}{k}\binom{2n-2k-1}{n} =$

$\dfrac{1}{2}n(n+1)$.

Proof Firstly, in the equality $(1+x)^{n+1} = \sum_{k=0}^{n+1} \binom{n+1}{k} x^k$, the

coefficient of x^{n-1} is equal to $\binom{n+1}{n-1} = \binom{n+1}{2} = \dfrac{1}{2}n(n+1)$. On the

other hand, in the equality

$$(1+x)^{n+1} = \frac{(1-x^2)^{n+1}}{(1-x)^{n+1}} = \left(\sum_{k=0}^{n+1}\binom{n+1}{k}(-1)^k x^{2k}\right)\left(\sum_{j=0}^{\infty}\binom{n+j}{n}x^j\right),$$

the coefficient of x^{n-1} is equal to

$$\sum_{k=1}^{[(n-1)/2]} (-1)^k \binom{n+1}{k}\binom{n+(n-1-2k)}{n}$$

$$= \sum_{k=1}^{[(n-1)/2]} (-1)^k \binom{n+1}{k}\binom{2n-2k-1}{n}.$$

Hence

$$\sum_{k=0}^{\lceil(n-1)/2\rceil} (-1)^k \binom{n+1}{k} \binom{2n-2k-1}{n} = \frac{1}{2}n(n+1).$$

Example 4   How many distinct ways are there to exchange n Yuan for 1 and 2 Yuan?

Solution   Suppose that there are a_n distinct ways, then $a_n = \sum_{t_1+2t_2=n} 1$ (where t_1, t_2 are nonnegative integers) and the generating function of a_n is

$$f(x) = \sum_{n=0}^{\infty} a_n x^n = \sum_{n=0}^{\infty} \Big(\sum_{t_1+2t_2=n} 1 \Big) x^n$$

$$= \sum_{t_1=0}^{\infty} \sum_{t_2=0}^{\infty} x^{t_1+2t_2} = \Big(\sum_{t_1=0}^{\infty} x^{t_1} \Big) \Big(\sum_{t_2=0}^{\infty} x^{2t_2} \Big)$$

$$= \frac{1}{1-x} \cdot \frac{1}{1-x^2} = \frac{1}{(1+x)(1-x)^2}$$

$$= \frac{A}{(1-x)^2} + \frac{B}{1-x} + \frac{C}{1+x}. \qquad\qquad ①$$

Thus $A = \frac{1}{1+x}\Big|_{x=1} = \frac{1}{2}$, $C = \frac{1}{(1-x)^2}\Big|_{x=-1} = \frac{1}{4}$. Setting $x=0$ in ①,

we get $A+B+C = 1$. Hence $B = 1-A-C = 1 - \frac{1}{2} - \frac{1}{4} = \frac{1}{4}$.

Therefore we have

$$f(x) = \sum_{n=0}^{\infty} a_n x^n = \frac{1}{2(1-x)^2} + \frac{1}{4(1-x)} + \frac{1}{4(1+x)}$$

$$= \frac{1}{2} \sum_{n=0}^{\infty} \binom{n+1}{1} x^n + \frac{1}{4} \sum_{n=0}^{\infty} x^n + \frac{1}{4} \sum_{n=0}^{\infty} (-1)^n x^n$$

$$= \frac{1}{4} \sum_{n=0}^{\infty} [2n+3+(-1)^n] x^n.$$

It follows that $a_n = \frac{1}{4}[2n+3+(-1)^n]$, i.e. there are $\frac{1}{4}[2n+3+(-1)^n]$ distinct ways.

Example 5(Example 8 in Chapter 1)   How many 3-digit integers are there such that the sum of digits of each integer is equal to 11?

Solution   We denote the hundred digit, ten digit and unit digit by the x_1, x_2, x_3 respectively, then

$$x_1 + x_2 + x_3 = 11 \ (1 \leqslant x_1 \leqslant 9, \ 0 \leqslant x_2 \leqslant 9, \ 0 \leqslant x_3 \leqslant 9). \quad ①$$

Thus the required number of 3-digit integers equals the number of integer solutions of the linear equation ①. But the number S of integer solutions of the linear equation ① equals the coefficient of x^{11} of the following function:

$$\begin{aligned}
f(x) &= (x + x^2 + \cdots + x^9)(1 + x + \cdots + x^9)^2 \\
&= (1 + x + \cdots + x^9)^3 - (1 + x + \cdots + x^9)^2 \\
&= \frac{(1 - x^{10})^3}{(1 - x)^3} - \frac{(1 - x^{10})^2}{(1 - x)^2} \\
&= (1 - 3x^{10} + 3x^{20} - x^{30}) \sum_{n=0}^{\infty} \binom{n+2}{2} x^n \\
&\quad - (1 - 2x^{10} + x^{20}) \sum_{n=0}^{\infty} \binom{n+1}{1} x^n.
\end{aligned}$$

So $S = \left(\binom{11+2}{2} - 3\binom{1+2}{2} \right) - \left(\binom{11+1}{1} - 2\binom{1+1}{1} \right) = 61.$ Therefore the required number of 3-digit integers is 61.

Exercise 3

1   Find the general term a_n of the following sequences of numbers using the generating functions:

(1) $a_0 = 2$, $a_1 = 5$, $a_{n+2} = 3a_{n+1} - 2a_n \ (n = 0, 1, 2, \ldots)$;

(2) $a_0 = 4$, $a_1 = 3$, $a_{n+2} = a_{n+1} + 6a_n - 12 \ (n = 1, 2, \ldots)$.

2   Prove that the following identities:

(1) $\left(\binom{n}{0} \right)^2 + \left(\binom{n}{1} \right)^2 + \cdots + \left(\binom{n}{n} \right)^2 = \binom{2n}{n}$;

(2) $\sum_{k=1}^{n} \binom{n}{k} \binom{n}{n+1-k} = \binom{2n}{n+1}$;

(3) $\displaystyle\sum_{k=0}^{[n/2]} (-1)^k \binom{n+1}{k} \binom{2n-2k}{n} = n+1;$

(4) $\displaystyle\sum_{k=0}^{n} (-1)^{n-k} 2^k \binom{n+k+1}{2k+1} = n+1;$

(5) $\displaystyle\sum_{k=p}^{n} (-1)^k \binom{n}{k} \binom{k}{p} = (-1)^n \delta_{pn}$, where $\delta_{ij} = \begin{cases} 1 & (i=j), \\ 0 & (i \neq j). \end{cases}$

Chapter 4   Recurrence Sequence of Numbers

4.1   Recurrence Sequence of Numbers

For a sequence $\{x_n\}$, if there is a positive integer k and an equation that connects x_{n+k} with its preceding k terms x_{n+k-1}, x_{n+k-2}, $\ldots$, x_n:

$$\Phi(x_{n+k}, x_{n+k-1}, \ldots, x_n) = 0, \qquad \text{①}$$

Then the sequence $\{x_n\}$ is called a k order recurrence sequence of numbers and the equation ① is called the recurrence relation of $\{x_n\}$. Solving x_{n+k} from the relation ①, we obtain

$$x_{n+k} = \varphi(x_{n+k-1}, x_{n+k-2}, \ldots, x_n). \qquad \text{②}$$

The equation ② is also called the recurrence relation of $\{x_n\}$. The first k terms of $\{x_n\}$:

$$x_1 = a_1, \, x_2 = a_2, \, \ldots, \, x_k = a_k \, (a_1, a_2, \ldots, a_k \text{ are given constants})$$
$$\text{③}$$

are called the initial values of $\{x_n\}$. Obviously, the sequence $\{x_n\}$ is determined uniquely by the recurrence relation ② and the initial values ③.

Suppose that the recurrence sequence $\{x_n\}$ is determined uniquely by the initial values ③ and the following recurrence relation:

$x_{n+k} = p_1 x_{n+k-1} + p_2 x_{n+k-2} + \cdots + p_k x_n + q$ (p_1, p_2, $\ldots$, p_k are constants and $p_k \neq 0$),

then the recurrence sequence $\{x_n\}$ is called the linear recurrence sequence of numbers (with constant coefficients) of order k. When

$q \equiv 0$, the recurrence sequence $\{x_n\}$ of numbers called homogeneous, otherwise it is non-homogeneous.

4.2 The Methods of Finding Solutions of Recurrence Relation

(1) **The method of characteristic roots**   Consider the linear homogeneous recurrence relation (with constant coefficients) of second-order:

$x_{n+2} = px_{n+1} + qx_n (n = 0, 1, 2, \ldots ; p, q$ are the constants and $q \neq 0)$.

$$\text{①}$$

If the geometric sequence $\{r^n\}$ $(r \neq 0)$ is a solution of the recurrence relation ①, we have $r^{n+2} = pr^{n+1} + qr^n$, i.e. r is a root of the following quadratic equation:

$$r^2 = pr + q. \qquad \text{②}$$

The quadratic equation ① is called the characteristic equation of the sequence $\{x_n\}$ and the roots of the equation ② are called the characteristic roots of the sequence $\{x_n\}$. Conversely, if r is a root of the equation ②, then the geometric sequence $\{r^n\}$ is a solution of ①.

If the two roots r_1 and r_2 of equation ② are distinct, then $\{r_1^n\}$ and $\{r_2^n\}$ are the solutions of ① and for any constants C_1 and C_2, $\{C_1 r_1^n + C_2 r_2^n\}$ also is the solution of ①.

If the initial values $x_1 = a$, $x_2 = b$ are given, the values of C_1 and C_2 are determined uniquely by the following system of equations:

$$\begin{cases} C_1 + C_2 = a, \\ C_1 r_1 + C_2 r_2 = b. \end{cases}$$

Therefore we get the unique solution $x_n = C_1 r_1^n + C_2 r_2^n$ of ① with the initial values $x_1 = a$, $x_2 = b$.

If $r = \frac{1}{2} p$ is a double root of equation ②, we get $pr + 2q = 0$ from the following system of equations:

$$\begin{cases} r = \dfrac{1}{2}p, \\ r^2 - pr - q = 0. \end{cases}$$

Thus

$$nr^n - p(n-1)r^{n-1} - q(n-2)r^{n-2}$$
$$= nr^{n-2}(r^2 - pr - q) + r^{n-2}(pr + 2q)$$
$$= 0.$$

So $\{nr^n\}$ is a solution of ① and for any constants C_1 and C_2, $\{C_1 r^n + C_2 nr^n\}$ is also the solution of ①. Since the values of C_1 and C_2 can be determined uniquely by the initial values $x_1 = a$, $x_2 = b$. Therefore, we get the unique solution $x_n = C_1 r_1^n + C_2 nr^n$ of ① with the initial values $x_1 = a$, $x_2 = b$.

Example 1 (**The Fibonacci Sequence of numbers**)   The rabbits mature in a month after birth. Each month the female of a mature pair gives birth to a new pair of rabbits with opposite sexes. At first, there is a pair of newly born rabbits with opposite sexes and how many pairs of rabbits are there at the end of the nth month?

Solution   Suppose that at the end of the nth month there are a_n pair of mature rabbits and b_n pair of newly born rabbits. If at the nth month there are F_n pair of rabbits, then $F_n = a_n + b_n$. But $a_n = a_{n-1} + b_{n-1} = F_{n-1}$ and $b_n = a_{n-1} = F_{n-2}$. Thus we have

$$F_n = a_n + b_n = F_{n-1} + F_{n-2}. \qquad\qquad ①$$

Obviously $F_0 = 1$ (Since we begin with only one pair of newly born rabbits.) and $F_1 = 1$ (Since at the end of first month, we only have one pair of mature rabbits.). The characteristic equation of ① is $r^2 = r + 1$. The characteristic roots are $r_1 = \dfrac{1+\sqrt{5}}{2}$, $r_2 = \dfrac{1-\sqrt{5}}{2}$. Thus

$$F_n = C_1 \left(\frac{1+\sqrt{5}}{2}\right)^n + C_2 \left(\frac{1-\sqrt{5}}{2}\right)^n.$$

Since $F_0 = 1$, $F_1 = 1$, we get

$$\begin{cases} C_1 + C_2 = 1, \\ C_1 \left(\dfrac{1 + \sqrt{5}}{2} \right) + C_2 \left(\dfrac{1 - \sqrt{5}}{2} \right) = 1. \end{cases}$$

Solving the system of equations, we get $C_1 = \dfrac{1 + \sqrt{5}}{2\sqrt{5}}$, $C_2 = -\dfrac{1 - \sqrt{5}}{2\sqrt{5}}$.

Hence

$$F_n = \frac{1}{\sqrt{5}} \left[\left(\frac{1 + \sqrt{5}}{2} \right)^{n+1} - \left(\frac{1 - \sqrt{5}}{2} \right)^{n+1} \right], \; n = 0, 1, 2, \ldots \quad \text{②}$$

Remark This sequence ② is called the Fibonacci sequence of numbers and its first 10 terms are

$$1, 1, 2, 3, 5, 8, 13, 21, 34, 55, \ldots$$

The problems concerning Fibonacci sequence of numbers often appear in mathematical competitions.

Example 2 How many n-digit numbers with no pair of adjacent 1's are there which consist of 1, 2 and 3?

Solution Assume that the number of required n-digit numbers is a_n. Obviously, $a_1 = 3$, $a_2 = 3^2 - 1 = 8$. We classify the n-digit numbers satisfying the conditions into two types: the number of n-digit numbers with the leading digit 1 and the second digit 2 or 3 is $2a_{n-2}$ and the number of n-digit numbers with the leading digit 2 or 3 is $2a_{n-1}$. By the addition principle, we get

$$a_n = 2a_{n-1} + 2a_{n-2} \, (n \geqslant 3). \quad \text{①}$$

The characteristic equation of ① is $r^2 = 2r + 2$. The characteristic roots are $r_1 = 1 + \sqrt{3}$, $r_2 = 1 - \sqrt{3}$. Thus

$$a_n = C_1 (1 + \sqrt{3})^n + C_2 (1 - \sqrt{3})^n.$$

Let a_0 satisfy $a_2 = 2a_1 + 2a_0$, i.e. $a_0 = \dfrac{a_2 - 2a_1}{2} = 1$. With $a_0 = 1$, $a_1 = 3$, we have

$$\begin{cases} C_1 + C_2 = 1, \\ C_1 (1 + \sqrt{3}) + C_2 (1 - \sqrt{3}) = 3. \end{cases}$$

Solving the above system of equations, we get

$$C_1 = \frac{2+\sqrt{3}}{2\sqrt{3}} = \frac{(1+\sqrt{3})^2}{4},$$

$$C_2 = -\frac{2-\sqrt{3}}{2\sqrt{3}} = -\frac{(1-\sqrt{3})^2}{4\sqrt{3}}.$$

Therefore, $a_n = \dfrac{1}{4\sqrt{3}}\left[(1+\sqrt{3})^{n+2} - (1-\sqrt{3})^{n+2}\right].$

(2) The Method of Substitution

In fact, it may be very difficult to find the solution of some recurrence relations. We will illustrate several examples about the method of substitution to obtaining the solution of some recurrence relations.

Example 3   Find a_n, if $a_1 = 1$, $a_n = \dfrac{2}{3}a_{n-1} + n^2 - 15$ $(n \geqslant 2)$.

Solution   Set undetermined constant a, b, c such that

$$a_n + (an^2 + bn + c) = \frac{2}{3}[a_{n-1} + (a(n-1)^2 + b(n-1) + c)],$$

i.e.

$$a_n = \frac{2}{3}a_{n-1} + \left(-\frac{1}{3}a\right)n^2 + \left(-\frac{4}{3}a - \frac{1}{3}b\right)n + \left(\frac{2}{3}a - \frac{2}{3}b - \frac{1}{3}c\right),$$

Comparing it with $a_n = \dfrac{2}{3}a_{n-1} + n^2 - 15$, we get

$$\begin{cases} -\dfrac{1}{3}a = 1, \\ -\dfrac{4}{3}a - \dfrac{1}{3}b = 0, \\ \dfrac{2}{3}a - \dfrac{2}{3}b - \dfrac{1}{3}c = -15. \end{cases} \Rightarrow \begin{cases} a = -3, \\ b = 12, \\ c = 15. \end{cases}$$

Thus

$$a_n + (-3n^2 + 12n + 15) = \frac{2}{3}[a_{n-1} + (-3(n-1)^2 + 12(n-1) + 15)].$$

Setting $b_n = a_n - 3n^2 + 12n + 15$, we have

$$b_n = \frac{2}{3}b_{n-1}, \ b_1 = a_1 - 3 + 12 + 15 = 25.$$

Thus $\{b_n\}$ is a geometric sequence with the leading term $a_1 = 25$ and the common ratio $q = \frac{2}{3}$. Hence we have $b_n = 25\left(\frac{2}{3}\right)^{n-1}$. Therefore

$$a_n = 25\left(\frac{2}{3}\right)^{n-1} + 3n^2 - 12n - 15.$$

Remark For the recurrence relation $a_n = pa_{n-1} + f(n)$, ($n \geqslant 2$, $p \neq 0$ is a constant and $f(n)$ is a polynomial of degree k). If $p \neq 1$, by applying the method of undetermined coefficient, we can determine the polynomial $g(n)$ of degree k such that

$$a_n + g(n) = p[a_{n-1} + g(n-1)].$$

Setting $b_n = a_n + g(n)$, we get $b_n = pb_{n-1}$, thus

$$b_n = b_1 p^{n-1} = [a_1 + g(1)]p^{n-1}.$$

It follows that

$$a_n = [a_1 + g(1)]p^{n-1} - g(n).$$

If $p = 1$, i.e. $a_n = a_{n-1} + f(n)$, using the method of iteration, we could obtain to get

$$\begin{aligned}
a_n &= a_{n-1} + f(n) \\
&= a_{n-2} + f(n-1) + f(n) \\
&= \cdots \\
&= a_1 + f(2) + f(3) + \cdots + f(n)
\end{aligned}$$

or

$$a_n = a_1 + \sum_{k=2}^{n}(a_k - a_{k-1}) = a_1 + \sum_{k=2}^{n} f(k).$$

Example 4 Find $a_1 a_2 \ldots a_n$, if $a_1 = 1$, $a_{n+1}a_n = 4(a_{n+1} - 1)$ $(n \geqslant 1)$.

Solution Since $a_{n+1}a_n = 4(a_{n+1} - 1)$, then $a_{n+1} = \dfrac{4}{4 - a_n}$,

$$a_{n+1} - 2 = \frac{4}{4 - a_n} - 2 = \frac{2(a_n - 2)}{4 - a_n}.$$

Thus

$$\frac{1}{a_n - 2} = \frac{4 - a_n}{2(a_n - 2)} = \frac{1}{a_n - 2} - \frac{1}{2}.$$

Hence $\left\{\dfrac{1}{a_n - 2}\right\}$ is a arithmetical sequence with the leading term $\dfrac{1}{a_1 - 2} =$

-1 and the common difference $d = -\dfrac{1}{2}$. Thus we get

$$\frac{1}{a_n - 2} = -1 + (n - 1)\left(\frac{-1}{2}\right) = -\frac{n+1}{2} \Rightarrow a_n = 2 \cdot \frac{n}{n+1}.$$

Therefore

$$a_1 a_2 \cdots a_n = \left(2 \cdot \frac{1}{2}\right)\left(2 \cdot \frac{2}{3}\right)\left(2 \cdot \frac{3}{4}\right)\cdots\left(2 \cdot \frac{n}{n+1}\right) = \frac{2^n}{n+1}.$$

Remark   In this example, 2 is the double roots of the equation $x^2 = 4(x - 1)$. More generally, for the fractional recurrence equation $a_{n+1} = \dfrac{aa_n + b}{ca_n + d}$, where $c \neq 0$, $ad - bc \neq 0$, $a_1 \neq \dfrac{aa_1 + b}{ca_1 + d}$. The roots of equation $x = \dfrac{ax + b}{cx + d}$ are called the fixed points of this sequence $\{a_n\}$ If the sequence $\{a_n\}$ has only one fixed point p (i.e. p is a double root of the equation $x(cx + d) = ax + b$). Then

$$\frac{1}{a_n - p} = \frac{1}{a_{n-1} - p} + \frac{2c}{a + d}.$$

If the sequence $\{a_n\}$ has two distinct fixed p and q, then ·

$$\frac{a_{n+1} - p}{a_{n+1} - q} = \frac{a - pc}{a - qc} \cdot \frac{a_{n-1} - p}{a_{n-1} - q}.$$

Example 5   Let $\dfrac{1}{2} < a_1 < \dfrac{2}{3}$, $a_{n+1} = a_n(2 - a_{n+1})$, $n = 1, 2, \ldots$.

Prove that

$$n + \frac{1}{2} < a_1 + a_2 + \cdots + a_n < n + 2.$$

Proof With $a_{n+1} = a_n(2 - a_{n+1})$, we get $a_{n+1} = \dfrac{2a_n}{a_n + 1}$. Solving the

equation $x = \dfrac{2x}{x + 1}$, we obtain two fixed points $x_1 = 0$, $x_2 = 1$. With

$a_{n+1} = \dfrac{2a_n}{a_n + 1}$ and $a_{n+1} - 1 = \dfrac{2a_n}{a_n + 1} - 1 = \dfrac{a_n - 1}{a_n + 1}$, we get $\dfrac{a_{n+1} - 1}{a_{n+1}} = \dfrac{1}{2} \cdot$

$\dfrac{a_n - 1}{a_n}$, i.e. $\left\{ \dfrac{a_n - 1}{a_n} \right\}$ is a geometric sequence with leading term $\dfrac{a_1 - 1}{a_1}$

and common ratio $\dfrac{1}{2}$. Hence

$$\frac{a_n - 1}{a_n} = \frac{a_1 - 1}{a_1} \left(\frac{1}{2} \right)^{n-1} \Rightarrow \frac{1}{a_n} = 1 + \left(\frac{1}{a_1} - 1 \right) \left(\frac{1}{2} \right)^{n-1}.$$

Since $\dfrac{1}{2} < a_1 < \dfrac{2}{3}$, $\dfrac{1}{2} < \dfrac{1}{a_1} - 1 < 1$,

$$1 + \frac{1}{2^n} < \frac{1}{a_n} < 1 + \frac{1}{2^{n-1}},$$

we get

$$\frac{1}{a_1} + \frac{1}{a_2} + \cdots + \frac{1}{a_n} < \left(1 + \frac{1}{1} \right) + \left(1 + \frac{1}{2} \right) + \cdots + \left(1 + \frac{1}{2^{n-1}} \right)$$

$$= n + 2 - \frac{1}{2^{n-1}} < n + 2$$

and

$$\frac{1}{a_1} + \frac{1}{a_2} + \cdots + \frac{1}{a_n} > \left(1 + \frac{1}{2} \right) + \left(1 + \frac{1}{2^2} \right) + \cdots + \left(1 + \frac{1}{2^n} \right)$$

$$= n + 1 - \frac{1}{2^n} > n + 1 - \frac{1}{2} = n + \frac{1}{2}.$$

This completes the proof.

Example 6 Find a_n, if $a_1 = 1$, $a_{n+1} = \dfrac{1}{16}(1 + 4a_n + \sqrt{1 + 24a_n})$ $(n \geqslant 1)$.

Solution For eliminating the radical, natuarally, we set $b_n = \sqrt{1 + 24a_n}$, i.e. $a_n = \dfrac{1}{24}(b_n^2 - 1)$. Substituting this to the original recurrence relation, we obtain

$$\frac{1}{24}(b_{n+1}^2 - 1) = \frac{1}{16}\left[1 + 4 \times \frac{1}{24}(b_n^2 - 1) + b_n\right],$$

i.e. $(2b_{n+1})^2 = (b_n + 3)^2$. Since $b_n > 0$, so

$$b_{n+1} = \frac{1}{2}(b_n + 3) \Rightarrow b_{n+1} - 3 = \frac{1}{2}(b_n - 3). \qquad \textcircled{1}$$

Thus $\{b_n - 3\}$ is a geometric sequence with the leading term

$$b_1 - 3 = \sqrt{1 + 24a_1} - 3 = 2$$

and the common ratio $q = \dfrac{1}{2}$. Hence

$$b_n - 3 = 2 \times \left(\frac{1}{2}\right)^{n-1} = 2^{2-n} \Rightarrow b_n = 2^{2-n} + 3.$$

Therefore

$$a_n = \frac{1}{24}(b_n^2 - 1) = \frac{1}{24}[(2^{2-n} + 3)^2 - 1] = \frac{1}{3}(2^{1-2n} + 3 \cdot 2^{-n} + 1).$$

Remark   In the equation $\textcircled{1}$, 3 is the root of equation $x = \dfrac{1}{2}(x + 3)$ and it is also called the fixed point of the function $f(x) = \dfrac{1}{2}(x + 3)$. Generally, the recurrence relation of degree 1 with the constant coefficients: $b_{n+1} = pb_n + q$ (p, q are the constants and $p \neq 1$) can be reduced to

$$b_{n+1} + \frac{q}{p - 1} = p\left(b_n + \frac{q}{p - 1}\right),$$

where $-\dfrac{q}{p - 1}$ is the root of the equation $x = px + q$.

Example 7   The sequences $\{a_n\}$, $\{b_n\}$ are defined as follows:

$$a_0 = \frac{\sqrt{2}}{2}, \ a_{n+1} = \frac{\sqrt{2}}{2}\sqrt{1 - \sqrt{1 - a_n^2}},$$

$$b_0 = 1, \ b_{n+1} = \frac{1}{b_n}(\sqrt{b_n^2 + 1} - 1), \ n = 0, 1, 2, \ldots.$$

Prove that the following inequality holds:

$$2^{n+2} a_n < \pi < 2^{n+2} b_n,$$

for each of $n = 0, 1, 2, \ldots$.

Proof Applying the mathematical induction, we easily get $0 < a_n < 1$, $b_n > 0$. Let $a_n = \sin \lambda_n \left(0 < \lambda_n < \dfrac{\pi}{2}, n \geqslant 0 \right)$, then

$$\sin \lambda_{n+1} = \frac{\sqrt{2}}{2} \sqrt{1 - \sqrt{1 - \sin^2 \lambda_n}} = \frac{\sqrt{2}}{2} \sqrt{1 - \cos \lambda_n} = \sin \frac{\lambda_n}{2}.$$

So $\lambda_{n+1} = \dfrac{1}{2} \lambda_n \, (n \geqslant 0)$ and $\lambda_0 = \arcsin \dfrac{\sqrt{2}}{2} = \dfrac{\pi}{4}$. Thus

$$\lambda_n = \frac{\pi}{4} \left(\frac{1}{2} \right)^n = \frac{\pi}{2^{n+2}} \Rightarrow a_n = \sin \lambda_n = \sin \frac{\pi}{2^{n+2}} \, (n \geqslant 0).$$

Similarly, let $b_n = \tan \delta_n$, we can obtain $b_n = \tan \dfrac{\pi}{2^{n+2}} \, (n \geqslant 0)$.

Since $\sin x < x < \tan x \left(0 < x < \dfrac{\pi}{2} \right)$, we get

$$a_n < \frac{\pi}{2^{n+2}} < b_n \Rightarrow 2^{n+2} a_n < \pi < 2^{n+2} b_n.$$

(3) The Mathematical Induction

The basic ideas to solve the problem about the recurrence relation of $\{x_n\}$ by applying the mathematical induction are that

(1) Use the recurrence relation and the initial values to compute the first several terms: $x_1, x_2, \ldots, x_k$ and explore the general law of $\{x_n\}$.

(2) Guess the expression of the general term of $\{x_n\}$ or some properties of $\{x_n\}$.

(3) Prove that the conjecture in (2) is true with the mathematical induction.

Example 8 Let a_n denote the number of the positive integers N, whose sum of digits is n and digits are 1, 3 or 4. Prove that for any $n = 1, 2, 3, \ldots$, a_{2n} is a perfect square. (China Mathematical

Competition in 1991)

Proof Among a_n positive integers satisfying the conditions, the number of numbers with leading digit 1, 3 or 4 are a_{n-1}, a_{n-3} or a_{n-4} respectively, then

$$a_n = a_{n-1} + a_{n-3} + a_{n-4} (n \geqslant 5) \qquad \textcircled{1}$$

Obviously $a_1 = 1$, $a_2 = 1$, $a_3 = 2$, $a_4 = 4$. Combining this with $\textcircled{1}$, we can evaluate the first several terms as follows:

n	1	2	3	4	5	6	7	8	9	10	11	12	$\ldots$
a_n	1	1	2	4	6	9	15	25	40	64	104	169	$\ldots$
law		1^2	1×2	2^2	2×3	3^2	3×5	5^2	5×8	8^2	8×13	13^2	

From the above table of numbers we guess that the following conclusion holds: Let $\{f_n\}$ be the Fibonacci sequence, i. e. $f_1 = 1$, $f_2 = 2$, $f_{n+2} = f_{n+1} + f_n (n \geqslant 1)$. Then the following conclusion holds:

$$\begin{cases} a_{2n} = f_n^2, \\ a_{2n+1} = f_n f_{n+1}, \end{cases} n = 1, 2, 3, \ldots. \qquad \textcircled{2}$$

Using the mathematical induction, we prove that for any positive integer n the above conclusion $\textcircled{2}$ is valid.

For $n = 1$ and $n = 2$, we have

$$a_2 = 1^2 = f_1^2, \ a_3 = 2 = 1 \times 2 = f_1 f_2,$$

and

$$a_4 = 2^2 = f_2^2, \ a_5 = 6 = 2 \times 3 = f_2 f_3,$$

i. e. for $n = 1$ and $n = 2$ the formula $\textcircled{2}$ is valid. We suppose that for $n = k - 1$ and $n = k$ the formula $\textcircled{2}$ is valid, then for $n = k + 1$, using the recurrence relation $\textcircled{1}$, the inductive hypothesis and the definition of the Fibonacci sequence, we get

$$\begin{aligned} a_{2(k+1)} &= a_{2k+1} + a_{2k-1} + a_{2k-2} = f_k f_{k+1} + f_{k-1} f_k + f_{k-1}^2 \\ &= f_k f_{k+1} + f_{k-1}(f_{k-1} + f_k) = f_k f_{k+1} + f_{k-1} f_{k+1} \\ &= f_{k+1}(f_k + f_{k-1}) = f_{k+1}^2, \end{aligned}$$

$$a_{2(k+1)+1} = a_{2(k+1)} + a_{2k} + a_{2k-1} = f_{k+1}^2 + f_k^2 + f_{k-1}f_k$$
$$= f_{k+1}^2 + f_k(f_k + f_{k-1}) = f_{k+1}^2 + f_k f_{k+1}$$
$$= f_{k+1}(f_{k+1} + f_k) = f_{k+1}f_{k+2},$$

i. e. for $n = k + 1$ our conjecture ② is true. There $a_{2n} = f_n^2$ is a perfect square for $n = 1, 2, 3, \ldots$. This completes proof.

Remark In this example, we also prove that

$$a_{2n} = (\sqrt{a_{2n-2}} + \sqrt{a_{2n-4}})^2 (n \geqslant 3) \qquad ③$$

is valid.

In fact, from ①, we can deduce the following recurrence relation:

$$a_{2n+4} = 2a_{2n+2} + 2a_{2n} - a_{2n-2}. \qquad ④$$

Using the mathematical induction and the recurrence relation ④, we could prove that the conjecture ③ is true. We leave the detail of the proof to the reader.

Example 9 Let the sequences $\{a_n\}, \{b_n\}$ satisfy the following conditions:

$$a_{n+1} = 7a_n + 6b_n - 3, \qquad ①$$

and

$$b_{n+1} = 8a_n + 7b_n - 4, \qquad ②$$

$n = 0, 1, 2, \ldots$. Prove that a_n is a perfect square for $n = 0, 1, 2, \ldots$. (China Mathematical Competition in 2000)

Proof I With ① we get

$$b_n = \frac{1}{6}(a_{n+1} - 7a_n + 3). \qquad ③$$

Substituting ③ to ② and rearranging we have

$$a_{n+2} = 14a_{n+1} - a_n - 6. \qquad ④$$

By direct computation, we get

$a_0 = 1 = 1^2$, $a_1 = 7a_0 + 6b_0 - 3 = 4 = 2^2$, $a_2 = 14a_1 - a_0 - 6 = 49 = 7^2$,
$a_3 = 14a_2 - a_1 - 6 = 676 = 26^2$, $a_4 = 14a_3 - a_2 - 6 = 9409 = 97^2$, $\ldots$.

and the sequence of numbers 1, 2, 7, 26, 97, ... satisfies the recurrence relation: $d_0 = 1$, $d_1 = 2$,

$$d_{n+2} = 4d_{n+1} - d_n (n = 0, 1, 2, \dots). \qquad \text{⑤}$$

Hence we guess that the formula $a_n = d_n^2$ is valid, where $\{d_n\}$ satisfies the recurrence relation ⑤. With the mathematical induction, we prove that this conjecture is true.

For $n = 0$ and $n = 1$, $a_0 = 1 = d_0^2$, $a_1 = 4 = d_1^2$. Assume that for $n = k - 1$ and $n = k$, we have that $a_{k-1} = d_{k-1}^2$, $a_k = d_k^2$. Then for $n = k + 1$, we get

$$\begin{aligned}
a_{k+1} &= 14a_k - a_{k-1} - 6 = 14d_k^2 - d_{k-1}^2 - 6 \\
&= (4d_k - d_{k-1})^2 - 2(d_k^2 + d_{k-1}^2 - 4d_k d_{k-1} + 3) \\
&= d_{k+1}^2 - 2(d_k^2 + d_{k-1}^2 - 4d_k d_{k-1} + 3).
\end{aligned}$$

But

$$\begin{aligned}
d_k^2 + d_{k-1}^2 - 4d_k d_{k-1} + 3 &= d_k(4d_{k-1} - d_{k-2}) + d_{k-1}^2 - 4d_k d_{k-1} + 3 \\
&= d_{k-1}^2 - d_{k-2}(4d_{k-1} - d_{k-2}) + 3 \\
&= d_{k-1}^2 + d_{k-2}^2 - 4d_{k-1}d_{k-2} + 3 \\
&= \cdots \\
&= d_1^2 + d_0^2 - 4d_1 d_0 + 3 \\
&= 2^2 + 1^2 - 4 \times 1 \times 2 + 3 = 0.
\end{aligned}$$

Thus $a_{k+1} = d_{k+1}^2$. Therefore for each of $n = 0, 1, 2, \dots$, $a_n = d_n^2$ is a perfect square.

Remark The recurrence relation ⑤ is also deduced by the method of undetermined coefficients.

In fact, setting $d_{n+2} = pd_{n+1} + qd_n$ and combining this with $d_0 = 1$, $d_1 = 2$, $d_2 = 7$, $d_3 = 26$, we get

$$\begin{cases} 2p + q = 7, \\ 7p + 2q = 26. \end{cases} \Rightarrow \begin{cases} p = 4, \\ q = -1. \end{cases}$$

Therefore $d_{n+2} = 4d_{n+1} - d_n (n = 0, 1, 2, \dots)$.

Proof Ⅱ Let $x_n = a_n - \dfrac{1}{2}$ ($\dfrac{1}{2}$ is the root of the equation $x = $

$14x - x - 6 \Big)$, then $x_0 = \dfrac{1}{2}$, $x_1 = \dfrac{7}{2}$ and from ④ we get

$$x_{n+2} = 14x_{n+1} - x_n \, (n \geqslant 0).$$

Using the method of the characteristic roots, we easily get

$$x_n = \frac{1}{4}(7 + 4\sqrt{3})^n + \frac{1}{4}(7 - 4\sqrt{3})^n$$

$$= \frac{1}{4}(2 + \sqrt{3})^{2n} + \frac{1}{4}(2 - \sqrt{3})^{2n}.$$

Thus

$$a_n = \frac{1}{4}(7 + 4\sqrt{3})^n + \frac{1}{4}(7 - 4\sqrt{3})^n + \frac{1}{2}$$

$$= \frac{1}{4}(2 + \sqrt{3})^{2n} + \frac{1}{4}(2 - \sqrt{3})^{2n} + \frac{1}{2}$$

$$= \left[\frac{1}{2}(2 + \sqrt{3})^n + \frac{1}{2}(2 - \sqrt{3})^n \right]^2.$$

Let $\dfrac{1}{2}(2 + \sqrt{3})^n = A_n + B_n \sqrt{3}$ (A_n, B_n are the positive integers), then

$$\frac{1}{2}(2 - \sqrt{3})^n = A_n - B_n \sqrt{3}.$$

Therefore $a_n = A_n^2$ is a perfect square.

Example 10 Set $N = (2 + \sqrt{3})^{2004}$. Find the first digit at the left of the decimal point and the first digit at the right of the decimal point of N.

Solution Set

$$x_n = (2 + \sqrt{3})^{2n} + (2 - \sqrt{3})^{2n} = (5 + 2\sqrt{6})^n + (5 - 2\sqrt{6})^n.$$

The characteristic equation is

$$[r - (5 + 2\sqrt{6})] \cdot [r - (5 - 2\sqrt{6})] = 0,$$

i.e. $r^2 - 10r + 1 = 0$. Thus the sequence of numbers $\{x_n\}$ satisfies the following recurrence relation:

$$x_{n+2} = 10x_{n+1} - x_n \, (n \geqslant 1). \hspace{2em} ①$$

Thus

$$x_1 = (5 + 2\sqrt{6}) + (5 - 2\sqrt{6}) = 10,$$
$$x_2 = (5 + 2\sqrt{6})^2 + (5 - 2\sqrt{6})^2 = 98$$

are two positive integers and $x_1 < x_2$. Suppose that x_{k-1}, x_k are two positive integers and $x_{k-1} < x_k$, and with ① we deduce that for each of $n = 1, 2, \ldots$, x_n is a positive integer and $x_{n-1} < x_n$. Hence

$$x_n = 10x_{n-1} - (10x_{n-3} - x_{n-4}) = 10(x_{n-1} - x_{n-3}) + x_{n-4}$$
$$\equiv x_{n-4} \pmod{10}$$

Especially we obtain $x_{1002} \equiv x_2 \equiv 8 \pmod{10}$, i. e. the unit digit of x_{1002} is 8. Since $0 < 5 - 2\sqrt{6} < 0.2$, thus

$$0 < (5 - 2\sqrt{6})^{1002} < 0.2^{1002} = 0.008^{334} < 0.01^{334} = \underbrace{0.00\cdots01}_{668 \text{ zeros}}$$

i. e.

$$x_{1002} = N + (5 - 2\sqrt{6})^{1002} = N + \underbrace{0.00\cdots0}_{669 \text{ zeros}}* * * \cdots.$$

Since the unit digit of x_{1002} is 8, therefore the first digit at the left of the decimal point of N is 7 and the first digit at the right of the decimal point of N is 9.

Exercise 4

1 Let the sequence $a_1, a_2, \ldots, a_n, \ldots$ of positive integers satisfy the following conditions:

(1) $\sqrt{a_n a_{n-2}} - \sqrt{a_{n-1} a_{n-2}} = 2a_{n-1}$; (2) $a_0 = a_1 = 1$.

Find a_n. (China Mathematical Competition in 1993)

2 Determine real number a_0 such that the sequence $\{a_n\}$ which is determined by the recurrence relation $a_{n+1} = -3a_n + 2^n$ ($n = 0, 1, 2, \ldots$) is increasing.

3 Let the sequence a_1, a_2, $\ldots$, a_n, $\ldots$ satisfy $a_1 = \dfrac{1}{2}$, $a_1 + a_2 + \cdots + a_n = n^2 a_n$, find a_n (7th Canadian Mathematical Olympiad in 1985).

4 Let $a_0 = 0$, $a_1 = 1$, $a_n = 2a_{n-1} + a_{n-2}$ ($n \geqslant 2$). Prove that $2^k \mid a_n$ if and only if $2^k \mid n$.

(Note: $a \mid b$ express that b is divisible by a). (The Problem Prepared for 29th IMO)

5 How many numbers of ways are there such that a $2 \times n$ chessboard can be perfect covered by n 1×2 rectangles without gaps and without overlaps?

6 Suppose a sphere is divided into a_n regions by n big circles on this sphere, no three of which are concurrent. Find a_n.

7 How many n-digit numbers can be formed by the digits 1, 2, 3, 4 such that the number of the digit 1 is even?

8 Four men A, B, C and D pass a ball to each other satisfying the following requirement: every man who accepts the ball pass this ball to any one of other three men at once. Suppose that A begin to pass the ball (as first time passing ball). In how many distinct ways can the ball return to A after 10 passes of the ball.

9 Prove that for any nonnegative integer n, the number $[(1 + \sqrt{3})^{2n+1}]$ is divisible by 2^{n+1}.

10 Prove that for any nonnegative integer n, the number $\sum\limits_{k=0}^{n} \dbinom{2n+1}{2k+1} 2^{3k}$ is not divisible by 35.

11 There is a sequence $\{a_n\}$ satisfying $a_0 = 1$, $a_{n+1} = \dfrac{7a_n + \sqrt{45a_n^2 - 36}}{2}$, $n \in \mathbf{N}$.

Prove that:

(1) For each $n \in \mathbf{N}$, a_n is a positive integer.

(2) For each $n \in \mathbf{N}$, $a_n a_{n+1} - 1$ is a perfect square.

(China Mathematical Competition in 2005)

Chapter 5 Classification and Method of Fractional Steps

5.1 Classification

When many situations are set in an investigative problem, we usually discuss each situations individually and find the solution. Summing and synthesizing these conclusions of the situations, we will obtain the solution of the original problem. That is the idea and method of classification.

When we solve some problems applying the idea and method of classification, the following rules must be followed:

(1) Each situation in the original problem must be contained in some class without omissions.

(2) Any two classes are disjoint and do not have overlaps.

(3) The classification must follow the same criterion.

(4) The key to choose the classifying criterion is that the problem appearing in each situation could be solved more easily than the original problem.

Example 1　How many 3-digit numbers can be formed from the digits 0, 1, 2, 3, 4, 5, 6, 7, 8, 9 such that the sum of its digits is even and greater than or equal to 10?

Solution　The 3-digite numbers whose sum of the digits is an even only have the two following classes: each digit is even or one is even and the other two are odd.

(1) When each digit is an even, the number of the 3-digit numbers is $P_3^5 - P_2^4 = 48$.

(2) When one digit is an even and the other two are odds, the

number of 3-digit numbers is

$$\binom{5}{1}\binom{5}{2}P_3^3 - \binom{5}{2}P_2^2 = 280.$$

But among above the 280 numbers, there are following 42 numbers with the sum of its digits less than 10:

204, 240, 402, 420; 206, 260, 602, 620; 103, 130, 301, 310;
105, 150, 501, 510; 107, 170, 701, 710; 305, 350, 503, 530;
123, 132, 213, 231, 312, 321; 125, 152, 215, 251, 512, 521;
134, 143, 314, 341, 413, 431.

Therefore the number of 3-digit numbers satisfying the conditions equals $48 + 280 - 42 = 286.$

Example 2 Suppose a figure consists of $2n$ $(n \geqslant 2)$ points (no four points of are coplanar) and there are $n^2 + 1$ line segments connecting these points. Prove that there exist n triangles in this figure.

Proof We shall prove the statement by induction on n. For the basic case $n = 2$, suppose the 4 points are A, B, C, D and there are $2^2 + 1 = 5$ line segments connecting these vertices.

Thus only $\binom{4}{2} - 5 = 1$ pair of these points is not connected by a line segment. Without loss of generality, suppose that there is no line segment connecting the points C and D. In this case, there exist two triangles: $\triangle ABC$ and $\triangle ABD$. Thus the basic case is proved.

Assume that the statement is true when $n = k \geqslant 2$. When $n = k + 1$, let a space figure G consist of $2k + 2$ points and there are $(k+1)^2 + 1$ line segments connecting these points. Firstly, we prove that there exists at least one triangle. Let the two given points A and B be connected by a line segment and the number of line segments connecting the points A (or B) and $2k$ other points is a (or b).

(1) If $a + b \geqslant 2k + 1$, then there are a point C (C differ from A and B) which is connected with A and B by the line segments. Thus there exists a triangle ABC.

(2) If $a + b \leqslant 2k$, by deleting the points A and B and all line segments meeting both A and B, then among residual figure, there are $2k$ points and at least

$$(k+1)^2 + 1 - 2k - 1 = k^2 + 1$$

line segments connecting these points. By the induction hypothesis, there exist k triangles.

Let $\triangle ABC$ be one of above triangles and n_A, n_B or n_C denote the number of line segments connecting the points A, B or C and other $2k - 1$ points respectively.

Case I: If $n_A + n_B + n_C \geqslant 3k - 1$, there are at least k triangles with one side of three sides AB, BC or CA respectively. Adding the triangle ABC, we have at least $k + 1$ triangles.

Case II: If $n_A + n_B + n_C \leqslant 3k - 2$, i. e.

$$(n_A + n_B) + (n_B + n_C) + (n_C + n_A) \leqslant 6k - 4,$$

then among three numbers $n_A + n_B$, $n_B + n_C$ and $n_C + n_A$, there is at least one is not exceeding $\left[\dfrac{6k-4}{3}\right] = 2k - 2$. Without loss of generality, assume $n_A + n_B \leqslant 2k - 2$. After deleting two points A and B and all line segments meeting both A and B, then among residual figure, there are $2k$ points and at least $(k+1)^2 + 1 - (2k-2) - 3 = k^2 + 1$ line segments connecting these points. By the induction hypothesis, there exist k triangles. Adding the triangle ABC, we have at least $k + 1$ triangles. Hence for $n = k + 1$, the conclusion is valid.

This completes the proof.

Remark   In this example, we have two classifications and the first classification and the second classification are independent (the criterions of two classifications are distinct). In the following example, our second classification is the subclasses of the first classification.

Example 3   Suppose 4 couples are seated on a bench to watch a movie such that any female is only adjacent to her husband or other female. How many different ways of seating are possible?

(The Problem Prepared for Japan Mathematical Olympiad in 1995)

Solution   Firstly, the number of ways of arranging 4 females is 4!. If there are the males between two females, then the number of the males is at least 2 (their husbands). Similarly, if there are females between two males, then there are at least 2 females. If seats of several females are consecutive, then we consider these females as a group. Thus unordered grouping ways of the females have the following 5 classes: 4, 3 + 1, 2 + 2, 2 + 1 + 1, 1 + 1 + 1 + 1. Since the isolated female must sit at two ends of the bench, the grouping way 1 + 1 + 1 + 1 does not satisfy the condition.

(1) When the grouping way of females is 2 + 1 + 1, the grouping way of males has only one class: 2 + 2. In this case, the mode of seating has on the only one class (The letters F and M express the female and the male respectively):

① F M M F F M M F

The number of way of arranging all males is only 1.

(2) When the grouping way of females is 2 + 2, the modes of seating have the 4 following classes:

② F F M M M M F F

③ M F F M M M F F or F F M M M F F M

④ M M F F M M F F or F F M M F F M M

⑤ M F F M M F F M

In the four classes above, the numbers of ways of arranging all males are 2!, 1, 1, and 1 respectively. The total ways is 2! + 1 × 2 + 1 × 2 + 1 = 7.

(3) When the grouping way of females is 3 + 1, the modes of seating have the 3 following classes:

⑥ F F F M M M M F or F M M M M F F F

⑦ M F F F M M M F or F M M M F F F M

⑧ M M F F F M M F or F M M F F F M M

In the three classes above, the numbers of ways of arranging all males are 2!, 1, and 1 respectively. The total ways is 2 × (2! + 1 + 1) = 8.

(4) When all of females are consecutive, the modes of seating

have the 3 following classes:

⑨ F F F F M M M M or M M M M F F F F

⑩ M F F F F M M M or M M M F F F F M

⑪ M M F F F F M M

In the three classes above, the numbers of ways of arranging all males are 3!, 2!, and 2! respectively. The total ways is

$$3! \times 2 + 2! \times 2 + 2! = 18.$$

Summarizing what we have described above, we obtain that the total ways of arranging 4 couples is

$$4! \times (1 + 7 + 8 + 18) = 24 \times 34 = 816.$$

Example 4 Suppose that a group of n ($n \geqslant 6$) persons satisfies the following conditions: (1) each person sends his regards to at most $n - \left[\dfrac{n+2}{2}\right]$ persons by telephone; (2) Among any three persons, there are at least two persons sending their regards to each other by telephone. Prove that it is possible to divide these n persons into two disjoint sets such that any two persons in the same set send their regards to each other by telephone.

Solution We represent a person with a point in the plane, no three points of which are collinear. If two persons send their regards to each other by telephone, then two corresponding points are connected by a red line segment, otherwise, connected by a blue line segment. Thus we obtain a figure G satisfying the following conditions: (1) There are at most $n - \left[\dfrac{n+2}{2}\right]$ red line segments meeting every point in G. (2) Among any three points in G, there are at least two points which are connected by a red line segment. Thus original problem is reduced to prove that it is possible to divide the n points in G into two disjoint subsets such that any two points in the same subset are connected by a red line segment.

Let S denote the set of n points in G. By the given condition (1), there are at least

$$n - 1 - \left(n - \left[\frac{n+2}{2} \right] \right) = \left[\frac{n}{2} \right]$$

blue line segments meeting every point of S. Suppose that the point A_1 and $[n/2]$ points B_1, B_2, ..., $B_{[n/2]}$ are connected by the blue line segments and the point B_1 and $[n/2]$ points A_1, A_2, ..., $A_{[n/2]}$ are connected by the blue line segments. Let

$$S_1 = \{A_1, A_2, \ldots, A_{[n/2]}\}$$

and

$$S_2 = \{B_1, B_2, \ldots, B_{[n/2]}\},$$

by the given condition (2), we know $S_1 \cap S_2 = \varnothing$ and any two points in the same subset $S_i (i = 1, 2)$ are connected by a red line segment.

Case I : When $n = 2k$ is even, then the set S is divided into two disjoint subsets S_1 and S_2 so that any two points in the same subset are connected by a red line segment. Thus the conclusion is valid.

Case II : When $n = 2k + 1 \geqslant 6$ is odd, then $k \geqslant 3$. Since $| S_1 \cup S_2 | = 2k$, then there is a point C in S which does not belong to $S_1 \cup S_2$.

① If C and B_1 are connected by a blue line segment, denote $S_1' = S_1 \cup \{C\}$ by the given condition (2), we know that any two points in each of S_1' and S_2 are connected by a red line segment and $S = S_1' \cup S_2$, $S_1' \cap S_2 = \varnothing$. Thus the conclusion is valid.

② If C and A_1 are connected by a blue line segment, with the same reason in ①, the conclusion is also valid.

③ If C and each of A_1 and B_1 are connected by the red line segments, with the given condition (1), there are at least $\left[\frac{n}{2} \right] = k \geqslant 3$ points and the point C are connected by the blue line segments and not all of these points belong to S_1 and not all of these points belong to S_2. Without loss of generality, suppose that C and a point $A_i (2 \leqslant i \leqslant k)$ in S_1 are connected by a blue line segment and C and the two points B_t, $B_j (2 \leqslant t < j \leqslant k)$ in S_2 are connected by two blue line segments. We know at least $\left[\frac{n}{2} \right] = k$ points and the point A_i are connected by the blue line segments and at least $k - 1$ of these points, except the point C, belong to S_2. But there are only $k - 2$ points in S_2, except two points B_t, B_j. So A_i and one of two points B_t, B_j must be connected by a blue line segment. For convenience, we assume that the point A_i and the point B_t is connected by a blue line segment. Thus any two of the three points C, A_i, B_t are connected by a blue segment, which contradicts the given condition (2). This completes the proof.

5. 2   The Method of Fractional Steps

The method of fractional steps is that the original complex and difficult problem is transformed into a group of correlative "small problems". Among these "small problems", the solutions of the posterior problems often depend on the solutions of the anterior problems. When the last problem is solved, we obtain the solution of the original problem.

Example 5   Find the smallest positive integer n such that among any n irrational numbers, there exist 3 irrational numbers such that the sum of any two of them is an irrational number, too.

Solution   Obviously, for the 4 following irrational numbers: $-\sqrt{2}$, $-\sqrt{3}$, $\sqrt{2}$, $\sqrt{3}$, any three of them contain two opposite numbers such that the sum of them equals zero and zero is not an irrational number. Hence the smallest positive number n satisfying the condition is greater than or equal to 5.

Next, we will prove that among any 5 irrational numbers, there are at least 3 numbers such that the sum of any two of them is an irrational number. For convenience, we represent 5 irrational numbers with 5 points x, y, z, u, v in the plane (The point and the corresponding irrational number are represented by the same letter) and not three of them are collinear. If the sum of two numbers is a rational number, the two corresponding points are connected by a red line segment, otherwise, by a blue line segment. Thus we obtain a figure G. and the original problem is reduced to prove that there exists a blue triangle in G.

Firstly, we prove that there exists a monochromatic triangle in G. If among 4 line segment meeting some point there are at least 3 red (or blue) line segments, then there must be a monochromatic triangle. Hence, without loss of generality, we could assume that there are exactly 2 red and 2 blue line segments meeting every points in G. And there are 5 red and 5 blue line segments in G. Since there are only two red line segments meeting every point in G, then every point is a vertex of a close broken line. But every close broken line has at least 3 line segments, hence 5 red line segments form a close broken line. Similarly, 5 blue line segments form a close broken line. For convenience, assume $xyzuvx$ is a red close broken line, thus $x + y$, $y + z$, $z + u$, $u + v$, $v + x$ are all rational numbers. Hence

$$x = \frac{1}{2}[(x + y) - (y + z) + (z + u) - (u + v) + (v + x)]$$

is a rational number, which contradicts the given condition. Consequently, there exists a monochromatic triangle.

Secondly, we prove that there exists a blue triangle. Otherwise, there exists a red triangle. For convenience, assume that $\triangle xyz$ is a red triangle. Thus $x + y$, $y + z$, $z + x$ all are rational numbers, and

$$x = \frac{1}{2}[(x + y) + (z + x) - (y + z)]$$

is a rational number, which contradicts the given condition. So there exists a blue triangle. Assume that $\triangle xyz$ is a blue triangle, i.e. there exist three irrational numbers x, y, z such that the sum of any two of them is an irrational number.

Summarizing what we have described above, we obtain that the smallest positive integer n is 5.

Example 6 Suppose that there are 2^{n-1} n-term sequences in which each term is 0 or 1. If for any three these sequences, there exists a positive integer p such that the pth terms of them all are 1. Show that there exists an unique positive integer k such that the kth terms of these 2^{n-1} sequences are all 1. (32 nd Moscow Mathematical Olympiad in 1969)

Proof Let $S = \{X \mid X = (x_1, x_2, \ldots, x_n), x_i = 0 \text{ or } 1, i = 1, 2, \ldots, n \}$ and S_0 represent the set of 2^{n-1} n-term sequences satisfying the given conditions. Thus S_0 is a proper subset of S.

Firstly, we discuss the characters of S_0 which are applied to prove the conclusion that we seek. For convenience, we need the following notes: for any $X = (x_1, x_2, \ldots, x_n)$, set $\overline{X} = (\overline{x_1}, \overline{x_2}, \ldots, \overline{x_n})$, where

$$\overline{x_i} = \begin{cases} 0, \text{ if } x_i = 1, \\ 1, \text{ if } x_i = 0, \end{cases} \quad i = 1, 2, \ldots, n,$$

and for any $X = (x_1, x_2, \ldots, x_n)$ and $Y = (y_1, y_2, \ldots, y_n)$, set

$$X \cdot Y = (x_1 y_1, x_2 y_2, \ldots, x_n y_n).$$

Thus $\overline{X}$ and $X \cdot Y \in S$, if $X, Y \in S$. With the given condition, for any $X, Y, Z \in S_0$, we know $X \cdot Y \cdot Z \neq (0, 0, \ldots, 0)$.

(1) We prove that for any $X \in S$, only one of the relations $X \in S_0$ and $\overline{X} \in S_0$ is valid and $0 = (0, 0, \ldots, 0) \notin S_0$.

Suppose that for any $X \in S$, X and $\overline{X}$ are paired. Thus the 2^n elements of S form 2^{n-1} pairs. If some X and $\overline{X}$ belong to S_0, then for any $Y \in S_0$, $X \cdot \overline{X} \cdot Y = (0, 0, \ldots, 0)$. This contradicts the given condition. Hence there are at most one of X, $\overline{X}$ belonging to S_0. Note that the number of the distinct pairs $(X, \overline{X})$ is 2^{n-1} and the number of sequences in S_0 is also 2^{n-1}, so there is exactly one of X, $\overline{X}$ belonging to S_0. Next, if $(0, 0, \ldots, 0) \in S_0$, then for any $X, Y \in S_0$ we know $0 \cdot X \cdot Y = (0, 0, \ldots, 0)$. This contradicts the given condition. Thus

$$0 = (0, 0, \ldots, 0) \notin S_0.$$

(2) Next we prove that if $X, Y \in S_0$, then $X \cdot Y \in S_0$. If $Z = X \cdot Y \notin S_0$, from (1), we obtain $\overline{Z} = \overline{Z} \cdot Y \in S_0$, then

$$X \cdot Y \cdot \overline{Z} = (X \cdot \overline{X}) \cdot (Y \cdot \overline{Y}) = (0, 0, \ldots, 0).$$

This contradicts the given condition. Thus $X \cdot Y \in S_0$.

(3) Finally, we prove that let X_1, X_2, ... , $X_{2^{n-1}}$ be 2^{n-1} distinct sequences in S_0 and $X = X_1 \cdot X_2 \cdot \cdots \cdot X_{2^{n-1}}$, then there is just one term in X which equals 1 and the remaining $n-1$ terms in X all equal zero.

From (1) and (2), we know $X \in S_0$ and $X \neq (0, 0, \ldots, 0)$, i. e. there are at lease one term of X which equals 1. If there are at least two terms of X which equal 1, the two terms of each sequence in S_0 equal 1 and the remaining $n-2$ terms of each sequence in S_0 all equal 1 or 0. Hence the number of the sequences in S_0 does not exceed 2^{n-2}, which contradicts that there are 2^{n-1} distinct sequences in S_0.

Secondly, we come back to the original problem. With (3), we could assume that the kth term of X is equal to 1 and the other terms of X are all equal to zero, i. e. there exists an unique positive integer k such that the kth terms of all sequences in S_0 are all equal to 1 and all the other terms of are not equal to 1. This completes the proof.

Remark　This problem is equivalent to the following problem. Let $I = \{a_1, a_2, \ldots, a_n\}$ be a set with n elements, and the set S_0 consist of 2^{n-1} distinct subsets of I satisfying the following conditions: the intersection of any three elements (i. e. three subsets of I) in S_0 is nonempty. Prove that the intersections of all elements (i. e. 2^{n-1} distinct subsets of I) in S_0 have an unique element. In the above solution, the operations $\overline{X}$ and $X \cdot Y$ just correspond to the operations of the complement and intersection of the sets. Hence the readers could use the perations of the complement and intersection of the sets to complete the proof.

Exercise 5

1　Suppose that 4-digit numbers $\overline{abcd}$ satisfy the following conditions: (1) $a, b, c, d \in \{1, 2, 3, 4\}$; (2) $a \neq b$, $b \neq c$, $c \neq d$, $d \neq a$; (3) a is the smallest number in a, b, c, d. Then number of

4-digit numbers $\overline{abcd}$ equals _____. (China Mathematical Competition in 2000)

2   The 5-digit numbers with at least three distinct digits are formed from the digits 1, 2, 3, 4, 5, 6. How many these 5-digit numbers are there such that the digits 1 and 6 are nonadjacent?

3   How many even integers with four different digits between 4000 and 7000 are there? (11st American Invitational Mathematical Examination in 1993)

4   There are 3 line segments whose lengths are 1, 2, 3 respectively and the line segment with length 3 is divided into any $n (\geqslant 2)$ line segments. Prove that among above $n + 2$ line segments, there must be 3 line segments which are three sides of a triangle.

5   Firstly, we select a subset X from set $S = \{1, 2, \ldots, n\}$. Next we select another subset Y of S such that X is not a subset of Y and Y is not a subset of X too. How many ways are there to select X and Y orderly?

6   In the rectangular coordinates system in plane, there are 9 integer points $A_i(x_i, y_i)$ $(x_i, y_i \in \mathbf{Z}, i = 1, 2, \ldots, 9)$, no three of which are collinear. Prove that there exists a $\triangle A_i A_j A_k (1 \leqslant i < j < k \leqslant 9)$ whose barycenter is an integer point too.

7   There are n people in a committee such that among any 3 people, there is a pair of mutually acquainted people. Find the smallest positive integer n such that there are 4 mutually acquainted people in these n persons?

8   There are 7 given points in a plane such that among any three points, at least two points are connected by a line segment. Find the smallest number of connected line segments and draw a figure satisfying above requirement. (The Problem Prepared for 30th IMO in 1989)

9   Suppose that the area of a convex hexagon $A_1 A_2 A_3 A_4 A_5 A_6$ equals S. Prove that there is at least a triangle $\triangle A_i A_j A_k (1 \leqslant i < j < k \leqslant 7)$ with area not exceeding $\frac{1}{6} S$.

Chapter 6   Correspondent Method

The correspondence is not only a basic mathematical concept but also an important method and skill to solve the problems and prove the propositions. What is more, it is a bridge between the unfamiliar and familiar problems. By a correspondence, covert relations are often transformed into overt relations such that we could find the way to solve the problems. When we solve some problems by the correspondent method, the key is to construct a corresponding relation. However, we do not have a general way to follow. We shall illustrate various correspondent methods to solve the problems in several examples.

6.1　Pairing Method

The pairing method is that with some rules, all objects of the investigative problem are paired so that the computation becomes simple and it is easy to find the solution.

Example 1　Let M be any k-element subset of set $I = \{1, 2, \ldots, n\}$. Assume that k elements of M are arranged in a decreasing order with $i_1 > i_2 > \cdots > i_k$ and $i_1 - i_2 + i_3 - \cdots + (-1)^{k-1}i_k$ is called the alternating sum of set M (For example, the alternating sum of set $\{1, 2, 4, 6, 9\}$ is $9 - 6 + 4 - 2 + 1 = 6$, and the alternating sum of set $\{5\}$ is 5). What is the total of all the alternative sums?

Solution　Let the subsets of I be divided into two class A and B such that each subset in A contain the number n and each subset in B does not contain number n and definite the alternating sum of empty set is zero. Let the set $\{a_1 < a_2 < \cdots < a_k < n\}$ in A and the set $\{a_1 < a_2 < \cdots < a_k\}$ in B be paired, then the number of such pairs is 2^{n-1} and the total of the alternating sums of any pairs: $\{a_1 < a_2 < \cdots < a_k < n\}$ and $\{a_1 < a_2 < \cdots < a_k\}$ is n. Therefore the total of all the alternating sums is $S = 2^{n-1} \cdot n$.

Example 2   There is a set $\dot{M} = \{1, 2, \ldots, 1000\}$. For any nonempty X of M, let α_X denote the sum of the greatest number and the smallest number in X. Then the arithmetic mean value of all α_X is equal to _____. (China Mathematical Competition in1991)

Solution   Let any nonempty subset X of M and $X' = \{1001 - x \mid x \in X\}$ be paired. Then X' is also a nonempty subset of M and $X' \neq X_1'$ if $X \neq X_1$. Thus all nonempty subsets of M are divided into two classes satisfying the following conditions: (1) $X = X'$; (2) $X \neq X'$. For $X = X'$, if $x \in X$, then $1001 - x \in X$, hence $1001 - x_0$ is the largest number of X, when x_0 is the smallest number of X. In this case,

$$\alpha_X = x_0 + (1001 - x_0) = 1001.$$

For $X \neq X'$, assume that the smallest number and the largest number of X are x_0 and y_0 respectively, then the smallest and largest numbers of X' are $1001 - y_0$ and $1001 - x_0$ respectively. In this case,

$$\alpha_X + \alpha_{X'} = x_0 + y_0 + (1001 - x_0) + (1001 - y_0) = 2002.$$

Summarizing what we have described above, we obtain the arithmetic mean value of all α_X is equal to 1001.

Example 3   Among any 133 positive integers, there are at least 799 pairs of coprime positive integers. Prove that there exist 4 distinct positive integers a, b, c, d such that a and b, b and c, c and d, d and a are relatively prime.

Proof   We represent 133 positive integers with 133 points A_1, A_2, $\ldots$, A_{133} in the plane. If two positive integers are relatively prime, then the two corresponding points are connected by a line segment, otherwise, not connected by a line segment. Thus we obtain a graph G satisfying the following condition: there are at least 799 edges connecting the 133 points in G. Hence the original problem is reduced to prove that there exists a quadrilateral with four vertices A, B, C, D and four edges AB, BC, CD, DA in G. Therefore we just need to prove that there exist two points B and D such that they are adjacent to both A and C. (Two points is called adjacent if they are connected by a line segment.)

If both of B and D are adjacent to the point A, then B and D are paired, and (B, D) is called the pair belonging to the point A. Assume that there are d_i edges meeting the point A_i ($i = 1, 2, \ldots, 133$), then

$$d_1 + d_2 + \cdots + d_{133} \geqslant 2 \times 799, \qquad \qquad ①$$

and the total of pairs belonging to each of A_1, A_2, $\ldots$, A_{133} is equal to

$$\sum_{i=1}^{133} \binom{d_i}{2} = \frac{1}{2} \left(\sum_{i=1}^{133} d_i^2 - \sum_{i=1}^{133} d_i \right). \qquad \qquad ②$$

By Cauchy's inequality, we obtain

$$\left(\sum_{i=1}^{133} d_i \right)^2 \leqslant \sum_{i=1}^{133} 1^2 \sum_{i=1}^{133} d_i^2,$$

i.e.

$$\sum_{i=1}^{133} d_i^2 \geqslant \frac{1}{133} \left(\sum_{i=1}^{133} d_i \right)^2. \qquad \qquad ③$$

Substituting ③ to ② and with ①, we know

$$\sum_{i=1}^{133} \binom{d_i}{2} \geqslant \frac{1}{2} \left[\frac{1}{133} \left(\sum_{i=1}^{133} d_i \right)^2 - \sum_{i=1}^{133} d_i \right]$$

$$= \frac{1}{2 \times 133} \left(\sum_{i=1}^{133} d_i \right) \left(\sum_{i=1}^{133} d_i - 133 \right)$$

$$= \frac{1}{2 \times 133} \times 2 \times 799 \times (2 \times 799 - 133)$$

$$> \frac{1}{2 \times 133} \times 2 \times 6 \times 133 \times (2 \times 6 \times 133 - 133)$$

$$= \frac{133 \times 132}{2} = \binom{133}{2}.$$

But the 133 points just form $\binom{133}{2}$ pairs, hence the above counting in left-hand side is repeated. So there is at least one pair (B, D) belonging to two distinct points A and C, i. e. the points B and D are adjacent to both A and C. Thus the four points A, B, C, D correspond to four positive integers a, b, c, d satisfying that a and b, b and c, c and d, d and a are relatively prime. This completes the proof.

Example 4 Let 11 sets M_1, M_2, $\ldots$, M_{11} satisfy the following conditions:

(1) $|M_i| = 5$, $i = 1, 2, \ldots, 11$;

(2) $|M_i \cap M_j| \neq 0$, $1 \leqslant i < j \leqslant 11$.

Denote $T = M_1 \cup M_2 \cup \cdots \cup M_{11}$, and for any $x \in T$, let

$$n(x) = |\{M_i \mid x \in M_i, 1 \leqslant i \leqslant 11\}|, \quad n = \text{Max}\{n(x) \mid x \in T\}.$$

Find the smallest value of n. (Romania Mathematical Olympiad in 1994)

Solution Let $T = \{x_1, x_2, \ldots, x_m\}$ and we construct an $m \times 11$ table such that the number in the unit square in the ith row and the jth column is

$$a_{ij} = \begin{cases} 1, & \text{if } x_i \in M_j, \\ 0, & \text{if } x_i \notin M_j, \end{cases} \quad (i = 1, 2, \ldots, m; j = 1, 2, \ldots, 11).$$

Thus $n(x_i) = \sum_{j=1}^{11} a_{ij}$ represents that x_i belongs to $n(x_i)$ sets in M_1, M_2, $\ldots$, M_{11}, where $n = \text{Max}_{1 \leqslant i \leqslant m}\{n(x_i)\}$ and $|M_j| = \sum_{i=1}^{m} a_{ij}$ represents the number of elements in set M_j. Hence from the given condition (1), we obtain

$$\sum_{i=1}^{m} n(x_i) = \sum_{i=1}^{m} \sum_{j=1}^{11} a_{ij} = \sum_{j=1}^{11} \sum_{i=1}^{m} a_{ij} = \sum_{j=1}^{11} |m_j| = 11 \times 5 = 55.$$

①

If x_k is a common element of sets M_i and $M_j (i < j)$, then (M_i, M_j) is called a pair of sets belonging to the element x_k. Thus the total of the pairs of sets belonging to each of x_1, x_2, ... , x_m is equal to

$$\sum_{i=1}^{m} \binom{n(x_i)}{2} = \frac{1}{2} \sum_{i=1}^{m} n(x_i)(n(x_i) - 1) \leqslant \frac{1}{2}(n-1) \sum_{i=1}^{m} n(x_i). \quad ②$$

In the other hand, from the given condition (2), for any (M_i, M_j) $(i < j)$, there exists a $y \in M_i \cap M_j$, i.e. (M_i, M_j) is a set belonging to y. Thus this set (M_i, M_j) is computed once at least on the left-hand side of ②. But there are $\binom{11}{2} = 55$ pairs of sets formed by M_1, M_2, ... , M_{11}, hence we know

$$55 \leqslant \sum_{i=1}^{m} \binom{n(x_i)}{2} \leqslant \frac{1}{2}(n-1) \sum_{i=1}^{m} n(x_i) \leqslant \frac{1}{2} \times 55 \times (n-1),$$

i.e. $n \geqslant 3$. If $n = 3$, then for any $x_i \in T$, $n(x_i) \leqslant n = 3$. Now we prove that there is no $x_i \in T$ such that $n(x_i) \leqslant 2$. Proof by contradiction, and assume that there is a $x_{i_0} \in T$, such that $n(x_{i_0}) \leqslant 2$, then

$$55 \leqslant \frac{1}{2} \sum_{i=1}^{m} n(x_i)(n(x_i) - 1)$$

$$= \frac{1}{2} \sum_{i \neq i_0} n(x_i)(n(x_i) - 1) + \frac{1}{2} n(x_{i_0})(n(x_{i_0}) - 1)$$

$$\leqslant \frac{1}{2}(n-1) \sum_{i \neq i_0} n(x_i) + \frac{1}{2} n(x_{i_0})(n(x_{i_0}) - 1)$$

$$= \frac{1}{2}(n-1) \sum_{i=1}^{m} n(x_i) + \frac{1}{2} n(x_{i_0})(n(x_{i_0}) - n)$$

$$\leqslant \frac{1}{2}(3-1) \times 55 + \frac{1}{2} \times 2 \times (2-3) = 54,$$

and this is a contradiction. Thus for any $x_i \in T$, $n(x_i) = 3 \Rightarrow 3m = 55$, which is impossible, so $n \geqslant 4$.

Finally, when $n = 4$, there exist 11 sets satisfying the given conditions (1) and (2):

$$M_1 = M_2 = \{1, 2, 3, 4, 5\}, M_3 = \{1, 6, 7, 8, 9\},$$
$$M_4 = \{1, 10, 11, 12, 13\}, M_5 = \{2, 6, 9, 10, 14\}$$
$$M_6 = \{3, 7, 11, 14, 15\}, M_7 = \{4, 8, 9, 12, 15\},$$
$$M_8 = \{5, 9, 13, 14, 15\}, M_9 = \{4, 5, 6, 11, 14\}$$
$$M_{10} = \{2, 7, 11, 12, 13\}, M_{11} = \{3, 6, 8, 10, 13\}.$$

Therefore $n(x_i) \leqslant 4$.

Summarizing what we have described above, we conclude that the smallest value of n is 4.

In the mathematical competitions, we apply not only the pairing method to the immediate count and the proof through the computation, but also apply it to solve a class of problems of two-person game. These problems often involve the characters of the geometric figures (for example, the symmetry and so on) and the properties of numbers (for example, the divisibility, the congruence and so on). In a problem of game, some step is called a "live step", if the game will not be ended if this step is performed. Hence, if we have a step which could be performed when the rival performs a "live step", then we shall never be defeated. But in a finite game, if one is not be defeated then he (or she) will win without fail (assume that the game does not have a tie).

The common winning strategy is that all possible selective positions (or all possible selective numbers) are paired such that each pair a, b satisfying that when the opponent chooses some position (or some number), say a, then you just need to choose another position b (or another number b).

In problems of game involving the geometric figure, the common pairing method is defined by the characters of the geometric figures (for example, the symmetry, the adjacency and so on) and the rules of game and we ought to occupy the symmetric center and the redundant positions in the figure. But in problems of games involving the properties of numbers, the common pairing method is defined by the characters of numbers (for example, the divisibility, the congruence and so on) and the rules of the game.

Example 5 On an 8×8 chessboard, two players A and B set the chessmen in turn and everytime one could set a chessman in several empty unit squares in the same row or in the same column (only one chessman in each gird). Assume that A goes first and then the players alternate. A player wins if another player could not set a chessmen any more. Who has a winning strategy? (21st Russian Mathematical Olympiad in 1995)

Solution We prove that the second player B has a winning strategy. Assume that at the kth step, the first player A sets m chessmen in m empty unit squares a_1, a_2, $\ldots$, a_m in the same row or in the same column, then the second player B just need to set m chessmen in m empty unit squares b_1, b_2, $\ldots$, b_m, where b_i and a_i are symmetrical about the center of the chessboard ($i = 1, 2, \ldots, m$). Therefore the second player B will win eventually.

Example 6 Alice and Bob play a game on a 6 by 6 grid. On his or her turn, a player chooses a rational number not yet appearing in the grid and writes it in an empty square of the grid. Alice goes first and then the players alternate. When we have written the numbers in all squares, in each row, the square with the greatest number in that row is colored black. Alice wins if she can then draw a line from the top of the grid to the bottom of the grid that stays in black squares, and Bob wins if she cannot. (If two squares share a vertex, Alice can draw a line from one to the other that stays in these two squares.) Find, with proof, a winning strategy for one of the plays. (33rd USA Mathematical Olympiad in 2006)

Solution Bob can win as follows. After each of his moves, Bob can ensure that the maximum number in each row is a square in $A \cup B$, where

$$A = \{(1, 1), (1, 2), (1, 3), (2, 1), (2, 2), (2, 3), (3, 1), (3, 2)\}$$

and

$$B = \{(3, 5), (4, 4), (4, 5), (4, 6), (5, 4), (5, 5),$$
$$(5, 6), (6, 4), (6, 5), (6, 6)\},$$

where (i, j) is the unit square in the ith row and the jth column.

In fact, Bob pairs each square of $A \cup B$ with a square in the same row that is not in $A \cup B$, so that each square of the grid is in exactly one pair. Whenever Alice plays in one square of a pair, Bob will play in the other square of the pair on his next turn. If Alice moves with x in $A \cup B$, Bob writes y with $y < x$ in the paired square. If Alice moves with x not in $A \cup B$, Bob writes z with $z > x$ in the paired square in $A \cup B$. So after Bob's turn, the maximum of each pair is in $A \cup B$ and thus the maximum of each row is in $A \cup B$.

So when all the numbers are written, the maximum square in the 6th row is in B and the maximum square in the 1st row is in A. Since there is no path from A to B that stays in $A \cup B$, Bob wins.

Example 7 Two players A and B play a game. The player A goes first and then the players alternate. Each time the player deletes any 9 numbers from the following 101 numbers: 1, 2, 3, ..., 101. After deleting numbers continually11 times, there are two numbers a and b $(a > b)$ which are not deleted, then the number of marks of A is $a - b$. Prove that player A can got at least 55 marks no matter how B goes. (32nd Moscow Mathematical Olympiad in 1969)

Proof Firstly, the player A deletes the following 9 numbers 47, 48, 49, ..., 55 and the remaining numbers are paired in 46 pairs $\{i, 55 + i\}$ $(i = 1, 2, ..., 46)$ such that the difference of two numbers in the same pair is 55. When the player B deletes 9 numbers every time, then the strategy of A is to delete the 9 numbers such that the 18 numbers deleted by A and B are paired in 9 pairs. Eventually, the two remaining numbers just are one of the above 46 pairs, so the difference of the two numbers equals 55. Therefore the player A could get at least 55 marks.

Example 8 On an infinite grid, two players A and B set chessmen in the empty unit square in turn. The player A set the black chessmen and the player B set the white chessmen. The player A goes first and then players alternate. The first player A wins if there are 11 black chessmen appearing in the continual unit squares at the same row or at the same column or at the same diagonal line. Prove that the second player B has a strategy such that the player A cannot win. (The Problem Prepared for (35th IMO in 1994).

Proof As in figure 6. 1: We write the digits 0, 1, 2, 3, 4 in each unit square in a 4×4 grid and a adjacent 2×2 grid. As in figure 6. 2, each figure containing two unit squares writing same digits is called a domino. Obviously, any 11 continual unit squares in the same row or in the same column or in the same diagonal line must contain a domino. If A set a black chessman in a unit square of some domino then B set a white chessman in another unit square of this domino, thus A can not wins. (When A sets a black chessman in a unit square writing the digit zero, then B sets a white chessman in another adjoining unite square writing the digit zero.)

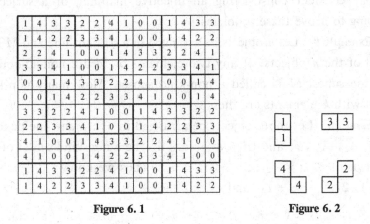

Figure 6. 1   Figure 6. 2

6. 2 Mapping Method

Using the mapping method to solve the problems, we often apply the following theorem.

Theorem   Let $f : M \rightarrow N$ be a mapping from the finite set M to the finite set N and $|M|$ and $|N|$ denote the number of the elements of M and N respectively.

(1) If f is an injective mapping (i. e. for any x, $y \in M$, if $x \neq y$, then $f(x) \neq f(y)$), then $|M| \leqslant |N|$;

(2) If f is a surjective mapping (i. e. for any $y \in N$, there exists $x \in M$, such that $f(x) = y$), then $|M| \geqslant |N|$;

(3) If f is a bijective mapping, or an one-to-one correspondence from M and N (i. e. f is an injective and surjective mapping) then $|M| = |N|$.

When it is difficult to count the number of a finite set M, we try to construct a bijective mapping between M and another finite set N and it is easy to count the number of the finite set N. Thus we obtain the number of the set M from $|M| = |N|$. This is the mapping method applied to counting problem. In some problems involving the combinatorial proof, we often must construct the inequalities. Then, we may consider constructing an injective mapping or a surjective mapping to prove these problems.

Example 9   Let n objects be arranged in a straight line and M be a subset of the n objects. If any two elements of M are non-adjoining, then the subset M is called a non-kind subset. How many non-kind subset with k elements are there?

Solution   Let the n objects be arranged in a straight line denoted by a_1, a_2, $\ldots$, a_n and $\{a_{i_1}, a_{i_2}, \ldots, a_{i_k}\}$ be a non-kind subset of the n objects $(1 \leqslant i_1 < i_2 < \cdots < i_k \leqslant n)$. Thus we know $i_{j+1} - i_j \geqslant 2$ $(j = 1, 2, \ldots, k-1)$, and

$$1 \leqslant i_1 < i_2 - 1 < i_3 - 2 < \cdots < i_k - (k-1) \leqslant n - (k-1).$$

Hence i_1, $i_2 - 1$, $i_3 - 2$, $\ldots$, $i_k - (k-1)$ is an increasing subsequence 1, 2, $\ldots$, $n - (k-1)$. Conversely, for every such increasing subsequence

$$1 \leqslant j_1 < j_2 < \cdots < j_k \leqslant n - (k-1),$$

the set containing the elements a_{j_1}, a_{j_2+1}, ... , $a_{j_k+(k+1)}$ is a non-kind set, thus we obtain an one-to-one correspondence between them. Since the number of ways to select the increasing subsequence j_1, j_2, ... , j_k from the sequence 1, 2, ... , $n - k + 1$ equals $\binom{n-k+1}{k}$, the number of the required non-kind subsets containing k elements equals $\binom{n-k+1}{k}$.

Example 10 Assume that no three diagonal lines intersect at the same interior point in a convex n-sided polygon. How many intersecting points of all diagonal lines in this polygon are there?

Solution Since an intersecting point P is determined by two diagonal lines l and m. Conversely, two intersecting diagonal lines l and m determine an intersecting point P. Therefore we may establish a one-to-one correspondence between P and (l, m). But two intersecting diagonal lines l and m are formed by connecting two vertex pairs (A, B) and (C, D) and for any four vertices of the convex n-sided polygon, there only exist two diagonal lines intersecting at an interior point in this polygon (as in figure 6.3). Hence we may establish an one-to-one correspondence between (l, m) and the four vertices group (A, B, C, D), i. e. $P \leftrightarrow (l, m) \leftrightarrow (A, B, C, D)$. Consequently there are as many intersecting points of the diagonal lines in the

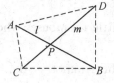

Figure 6.3

convex n-sided polygon as there are the four vertex groups formed by the n vertices of the polygon, i. e. $\binom{n}{4}$.

Remark The result of this problem is an important conclusion of the combinatorial geometry which is often applied to solve more difficult counting problems of the combinatorial geometry.

Example 11 Consider an equilateral triangle ABC with side length n, which is divided perfectly into little triangles with unit side length, as shown in figure 6.4. Find the number of the little rhombuses with unit side length in the figure.

Solution   Firstly, we consider the little rhombuses whose sides are not parallel with BC. Extend the four sides of each little rhombus and intersect BC at four dividing points. (In an especial case, the second intersecting point and third intersecting point coincide with a vertex of this rhombus.) For the convenience of computation, we extend AB

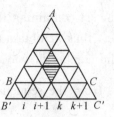

Figure 6. 4

and AC to the points B' and C' respectively such that $BB' = CC' = 1$ and extend all parallel lines which intersect line segment $B'C'$ at $n + 2$ points of division, denoted by $0, 1, 2, \ldots, n + 1$ (containing two end points B' and C'). Thus each little rhombus whose sides are not parallel with BC corresponds to four different points of division i, $i + 1$, k, $k + 1$ on the line segment $B'C'$. Conversely, any four different points on the line segment $B'C'$ correspond to a little rhombus whose sides are not parallel with BC. It is a one-to-one correspondence, and there exists a one-to-one correspondence between the order groups $(i, i+1, k, k+1)$ $(0 \leqslant i < i+1 < k < k+1 \leqslant n+1)$ and the order groups $(i+1, k)$ $(1 \leqslant i+1 < k \leqslant n)$ too. Consequently, the number of the little rhombuses whose sides are not parallel with BC equals $\binom{n}{2}$. By the symmetry, the required number of the little rhombuses is $3\binom{n}{2}$.

Example 12   Let S be a finite set of points in a space with an orthogonal system of coordinates $Oxyz$. Let S_x, S_y, S_z be the sets of orthogonal projection of S on the planes Oyz, Ozx, Oxy respectively. Prove that

$$|S|^2 \leqslant |S_x| \cdot |S_y| \cdot |S_z|,$$

where $|A|$ denotes the number of elements in the finite set A. (33rd IMO)

Solution   For any $(i, j, 0) \in S_z$, let $T_{ij} = \{(i, j, z) \mid (i, j, z) \in S\}$. Obviously we know $S = \bigcup_{(i, j, 0) \in S_z} T_{ij}$. By Cauchy's inequality, we obtain

$$| S | \leqslant \sum_{(i,j,0) \in S_z} 1^2 \cdot \sum_{(i,j,0) \in S_z} | T_{ij} |^2 = | S_z | \sum_{(i,j,0) \in S_z} | T_{ij} |^2. \quad \textcircled{1}$$

Consider the set $V = \bigcup_{(i,j,0) \in S_z} (T_{ij} \times T_{ij})$, where

$$T_{ij} \times T_{ij} = \{((i,j,t_1),(i,j,t_2)) \mid (i,j,t_k) \in S, k = 1, 2\},$$

thus

$$| V | = \sum_{(i,j,0) \in S_z} | T_{ij} |^2.$$

We definite the mapping $f : V \to S_x \times S_y$ as follows:

$$V \ni ((i,j,t_1), (i,j,t_2)) \to ((0,j,t_1), (i,0,t_2)) \in S_x \times S_y.$$

We know easily that f is an injection, hence

$$| V | \leqslant | S_x \times S_y | = | S_x | \cdot | S_y |. \quad \textcircled{2}$$

From $\textcircled{1}$ and $\textcircled{2}$, we obtain

$$| S |^2 \leqslant | S_x | \cdot | S_y | \cdot | S_z |.$$

This completes the proof.

Example 13 Assume that the family of sets Ω consists of some ternary subsets of $I = \{1, 2, \ldots, n\}$ such that any two elements of Ω (i.e. two ternary subsets of I) have one common element at most. Prove that there exists a subset X of I satisfying the following conditions: (1) Any element of Ω (i.e., a ternary subset of I) is not the subset of X; (2) $| X | \geqslant [\sqrt{2n}]$.

Analysis We just need to find a subset X of I satisfying the condition (1) such that the number of the elements of X is maximal and the inequality $| X | \geqslant [\sqrt{2n}]$ holds.

Proof A subset M of I is called a good subset if any element of Ω (i.e. a ternary subset of I) is not the subset of M. Obviously the good subset is existent. (Since any binary subsets of I are all good subsets) and the number of the good subsets is finite. Suppose that the subset X of I is a good subset whose number of elements is the most and $| X | =$

k, then X satisfies the condition (1). Hence we just need to prove that X satisfies the condition (2).

Denote $Y = I \backslash X$, from the given condition (1), we know $X \neq I$. It follows $Y \neq \varnothing$. Let $Y = \{y_1, y_2, \ldots, y_{n-k}\}$, since the number of the elements of X is maximal, then for any $y_i \in Y$, the set $X \cup \{y_i\}$ is not a good subset, i. e. there exists a element A_i of Ω (i. e. a ternary subsets of I) such that $A_i \subset X \cup \{y_i\}$. But $A_i \not\subset X$, thus there exist two elements x_{i_1}, $x_{i_2} \in X$ such that $A_i = \{x_{i_1}, x_{i_2}, y_i\}$. We define the mapping $f : Y \to \mathfrak{J}$ as follows: $Y \ni y_i \to \{x_{i_1}, x_{i_2}\} \in \mathfrak{J}$, where the family of sets $\mathfrak{J}$ consists of all binary subsets of X, then f is an injective mapping. In fact, for any y_i, $y_j \in Y$, $y_i \neq y_j$, there exist two binary subsets $\{x_{i_1}, x_{i_2}\}, \{x_{j_1}, x_{j_2}\} \in \mathfrak{J}$ such that $A_i = \{x_{i_1}, x_{i_2}, y_i\}$, $A_j = \{x_{j_1}, x_{j_2}, y_j\} \in \Omega$. If $f(y_i) = f(y_j)$, i. e. $\{x_{i_1}, x_{i_2}\} = \{x_{j_1}, x_{j_2}\}$, then the two ternary subsets A_i and A_j have two common elements, which contradicts the given condition. Therefore f is an injective mapping. It follows that $|Y| \leqslant |\mathfrak{J}|$, i. e. $n - k \leqslant \binom{k}{2}$, $[\sqrt{k(k+1)}] \geqslant [\sqrt{2n}]$. But

$$k < \sqrt{k(k+1)} < k + 1,$$

thus

$$|X| = k = [\sqrt{k(k+1)}] \geqslant [\sqrt{2n}].$$

Example 14 There are 40 persons at a conference. Every 19 persons share exactly one common idol at the conference. (If A is an idol of B, then B is not necessarily the idol of A, Also A is not his own idol). Prove that at this conference, there exists a set T_0 containing 20 persons such that for any $P \in T_0$, P is not the common idol of the remaining 19 persons in T_0.

Proof Let S denote the set of 40 persons at this conference. A 20-element subset T of S is called a good subset, if there is one person $P \in T$ such that the unique common idol of the remaining 19 persons is P. Hence we just need to prove that there are at least one 20-element subset of S which is not a good subset. It is equivalent to prove that the number of the good subsets smaller than $\binom{40}{20}$, the number of all 20-element subsets of S.

Let X denote the family of sets consisting of all good subsets of S and Y denote the family of sets consisting of all 19-element subsets of S. Moreover, for any $B \in Y$, we use $g(B)$ to denote the unique common idol of 19 persons in B. From the definition of the good subset we know that for any $T \in X$, there is a person $P \in T$ such that the unique common idol of the remaining 19 persons is P, i. e. $g(T\backslash\{P\}) = P$. Thus we could definite a mapping $f : X \to Y$ as follows:

$$X \ni T \to T\backslash\{P\}.$$

Then, we prove that f is an injective mapping. Indeed, for any T_1, $T_2 \in X$, $T_1 \neq T_2$, there are $P_1 \in T_1$, $P_2 \in T_2$ such that $g(T_1\backslash\{P_1\}) = P_1$, $g(T_2\backslash\{P_2\}) = P_2$ and $f(T_1) = T_1\backslash\{P_1\}$, $f(T_2) = T_2\backslash\{P_2\}$. If $f(T_1) = f(T_2)$, i.e. $T_1\backslash\{P_1\} = T_2\backslash\{P_2\}$. Thus

$$P_1 = g(T_1\backslash\{P_1\}) = g(T_2\backslash\{P_2\}) = P_2,$$

and it is a contradiction. Thus f is an injective mapping. Consequently, we obtain that

$$|X| \leqslant |Y| = \binom{40}{19} < \binom{40}{20}.$$

Since the number of 20-element subsets of S equals $\binom{40}{20}$, there is at least one 20-element subset T_0 of S which is not a good subset. Therefore for any $P \in T_0$, the unique common idol of other 19 persons is not P. This completes the proof.

Example 15 Let the set M consist of 48 distinct positive integers whose prime divisors are smaller than or equal to 30. Prove that there are four distinct positive integers of M such that their product is a perfect square. (49th Moscow Mathematical Olympiad in 1986)

Proof There are 10 primes not exceeding 30: $p_1 = 2$, $p_2 = 3$, $p_3 = 5$, $p_4 = 7$, $p_5 = 11$, $p_6 = 13$, $p_7 = 17$, $p_8 = 19$, $p_9 = 23$, $p_{10} = 29$. Let

$$Y = \{p_1^{\alpha_1} p_2^{\alpha_2} \cdots p_{10}^{\alpha_{10}} \mid \alpha_i = 0 \text{ or } 1, i = 1, 2, \ldots, 10\}$$

and X be the family of sets consisting of all binary subsets of $M.$.

For any $\{a, b\} \in X$, the product ab could be written uniquely in the following form: $ab = K_{ab}^2 \cdot m_{ab}$, where $K_{ab} \in \mathbf{N}_+$, $m_{ab} \in Y$. Let $\{a, b\}$ correspond to the positive integer m_{ab}, thus this correspondence forms a mapping f from X to Y. Since $|X| = \binom{48}{2} = 1128$, $|Y| = 2^{10} = 1024$, $|X| > |Y|$, f is not an injective mapping. Therefore there are $\{a, b\}, \{c, d\} \in X, \{a, b\} \neq \{c, d\}$ such that

$$m_{ab} = f(\{a, b\}) = f(\{c, d,\}) = m_{cd}.$$

It follows that

$$abcd = K_{ab}^2 m_{ab} \cdot K_{cd}^2 m_{cd} = (K_{ab} K_{cd} m_{ab})^2$$

is a perfect square. If a, b, c, d are distinct, then the conclusion holds. Otherwise, there are exactly two equal numbers in $\{a, b\}$ and $\{c, d\}$. Without loss of generality, assume $a \neq c$, and $b = d$, as $abcd = acb^2$ is a perfect square, we know ac is also a perfect square.

Since $|M\backslash\{a, b\}| = 46$, and $\binom{46}{2} = 1035 > 1024 = |Y|$, in a similar way, there are two distinct binary subsets $\{a', b'\}$ and $\{c', d'\}$ of the family $X\backslash\{\{c, d\}\}$ such that the product $a'b'c'd'$ is a perfect square. If a', b', c', d' are distinct, then the conclusion holds. Otherwise, in a similar way, we may assume that $a' \neq c'$, $b' = d'$ and $a'c'$ is a perfect square. Consequently, there exist four disjoint numbers a, c, a', $c' \in M$ such that $aca'c'$ is a perfect square. This completes the proof.

Besides the correspondent method of one-to-one, we also use the correspondent method of one-to-many.

Example 16 Assume that there are n $(n \geq 6)$ points on a circle, and each pair of points is connected by a line segment, and any three line segments do not intersect at the same interior point of this circle. Thus any three line segments which intersect at three points determine a triangle. Find the number of triangles which are determined by these line segments. (The Training Problem for National Team of China for 32^{sd} IMO)

Solution　Suppose that the points on the circle are called the exterior points, the intersecting points of any two line segments in the circle are called the interior points. Thus these triangles are divided into four groups:

(1) In the first group of the triangles, the three vertices of every triangle are all the exterior points. Obviously, any three points on the circle correspond to just a triangle belonging to the first class. Therefore the number of first class of triangles is $\binom{n}{3}$.

(2) In the second group of triangles, the two vertices of each triangle are the exterior points and the third vertex is a interior point. In this case, as the figure 6.5(1), any four points A_1, A_2, A_3, A_4 on the circle correspond to 4 triangles $\triangle A_1OA_2$, $\triangle A_2OA_3$, $\triangle A_3OA_4$, $\triangle A_4OA_1$ belonging to the second group. Therefore the number of second group of triangles is $4\binom{n}{4}$.

(3) In the third group of triangles, the two vertices of every triangle are the interior points and the third vertex is an exterior point. In this case, as in figure 6.5(2), any five points A_1, A_2, A_3, A_4, A_5 on the circle correspond to 5 triangles $\triangle A_1B_1B_2$, $\triangle A_2B_2B_3$,

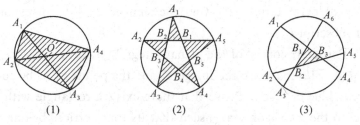

(1)　　　　　(2)　　　　　(3)

Figure 6.5

$\triangle A_3 B_3 B_4$, $\triangle A_4 B_4 B_5$, $\triangle A_5 B_5 B_1$ belonging to the third class. Therefore the number of the third class of the triangles is $5\binom{n}{5}$.

(4) In the fourth group of triangles, the three vertices of every triangle are all the interior points. In this case, as the figure 6.5(3), any six points A_1, A_2, A_3, A_4, A_5, A_6 on the circle correspond to a triangle $\triangle B_1 B_2 B_3$ belonging to the forth group. Therefore the number of the fourth group of the triangles is $\binom{n}{6}$.

Summarizing what we have described above, the required number of the triangles is

$$\binom{n}{3} + 4\binom{n}{4} + 5\binom{n}{5} + \binom{n}{6}.$$

Exercise 6

1 Let $S = \{1, 2, \ldots, 100\}$. For any 55-element subset M of S, are there two numbers belonging to M such that their difference equals (1) 9; (2) 11?

2 Let $n \equiv 1 \pmod 4$ and $n > 1$, and $P = \{a_1, a_2, \ldots, a_n\}$ be any permutation of $\{1, 2, \ldots, n\}$, and let k_p denote the largest subscript k satisfying the following condition:

$$a_1 + a_2 + \cdots + a_k < a_{k+1} + a_{k+2} + \cdots + a_n.$$

For all permutations P, find the sum of all the corresponding largest subscript k_p.

3 Let set M consist of the planar integral points (x, y) ($1 \leqslant x \leqslant 12$, $1 \leqslant y \leqslant 12$, x, y are the integers) and the points of M be colored with red, blue or black. Prove that there exists a rectangle with sides parallel to the x-axis or y-axis such that its four vertices belong to M and have the same color.

4 Set a chessman in a cell of an $m \times n$ chessboard. Two players move this chessman in turn with the following rule: Every time, one may move the chessman from one cell to another adjacent cell (if two cells have the common side, then they are adjacent), but one did not move it to the cell which it had arrived at. The player is defeated if he (or she) cannot move the chessman.

(1) Suppose that the initial position of the chessman is the lower left cell. Who has a winning strategy?

(2) Suppose that the initial position of the chessman is the adjacent cell of the lower left cell. Who has a winning strategy?

5 Assume that we write the digits $1, 2, \ldots, n$ in order in n unit squares of $1 \times n$ rectangle. Set three chessmen in three unit squares which are written with digits $n - 2$, $n - 1$, n such that each of three unit squares contains a chessman. The two players A and B play the game in turn. The player A goes first and then the players alternate. At every step, the player may moves one of three chessmen from one unit square to another empty unit square in which the writing digit is less. If one cannot move the chessman, then he (or she) is defeated. Prove that player A has a winning strategy.

6 Consider an equilateral triangle ABC of side length n, which is divided into the equilateral triangles with the unit side length, as show in figure 6. 4. Let $f(n)$ denote the number of the parallelograms in the figure. Find $f(n)$.

7 Let $S = \{1, 2, \ldots, 1000\}$ and A is a 201-element subset of S. If the sum of the elements of A is divisible by 5, then subset A is called a good subset. Find the number of all good subsets of S.

8 In a math camp, any m students share exactly one common friend. $m \geqslant 3$. (If A is a friend of B, then B is a friend of A. Also, A is not his own friend.) Suppose a person P has the largest number of friends. Determine what that number is. (China National Team Selection Test for 31st IMO in 1990)

9 Let A_i be the finite set, $i = 1, 2, \ldots, n$. Prove that if

$\displaystyle\sum_{1\leqslant i<j\leqslant n} \frac{|A_i \cap A_j|}{|A_i|\cdot|A_j|} < 1$, then there exist $a_i \in A_i$, $(i = 1, 2, \ldots, n)$ such that $a_i \neq a_j$ if $i \neq j$ $(1 \leqslant i < j \leqslant n)$.

10 Prove that every sequence a_1, a_2, $\ldots$, a_{mn+1} of $nm + 1$ real numbers contains an increasing subsequence of length $n + 1$ or a decreasing subsequence of length $m + 1$. (The length of a sequence is the number of terms of this sequence.)

11 Suppose the sequences of length 15 consists of two letters a and b. How many different sequences will contain exactly five aa, three ab, three ba, and three bb subsequences in above sequences?

12 Assume that n is even. How many ways are there such that we can select 4 distinct numbers a, b, c, d from the set $\{1, 2, \ldots, n\}$ satisfying $a + c = b + d$? (a, b, c, d are unordered.)

13 An n-term sequence $(x_1, x_2, \ldots, x_n)$ in which each term is either 0 or 1 is called a binary sequence of length n. Let a_n be the number of binary sequences of length n containing no three consecutive terms equal to 0, 1, 0 in that order. Let b_n be the number of binary sequences of length n that contain no four consecutive terms equal to 0, 0, 1, 1 or 1, 1, 0, 0 in that order. Prove that $b_{n+1} = 2a_n$ for all positive integers n. (25th USA Mathematical Olympiad in 1996)

Chapter 7 Counting in Two Ways

Let $A = \{a_1, a_2, \dots, a_m\}$, $B = \{b_1, b_2, \dots, b_n\}$ be two finite sets. The Cartesian product of the two sets A and B is the set of ordered pairs with the form (a_i, b_j), i.e.

$$A \times B = \{(a_i, b_j) \mid 1 \leqslant i \leqslant m, 1 \leqslant j \leqslant n\}.$$

Suppose that for any $a_i \in A$, $C_i = \{(a_i, b) \mid b \in B\}, (i = 1, 2, \dots, m)$ and for any $b_j \in B$, $D_j = \{(a, b_j) \mid a \in a\}$, $(j = 1, 2, \dots, n)$, then

$$|A \times B| = \sum_{i=1}^{m} |C_i| = \sum_{j=1}^{n} |D_j|.$$

This equality is called the Fubini's principle or the principle of counting in two ways.

When solving some problems referring to the computation of the quantitative relation, we usually use two different ways to count the objects and synthesize the computing results which are applied to find the answers of these problems. It is the principle of counting in two ways.

Example 1 There are 1000 points in a square, and among these points and the vertices of the square, no three points are collinear. These points and the vertices of the square are connected by the line segments such that the square is divided perfectly into several triangles. (The sides of these triangles are all the connecting line segments and the sides of the original square, and any two connecting line segments have no common interior point except the end points). Find the number of the connecting line segments and the number of the triangles in this figure.

Solution   Assume that there are l line segments and k triangles in this figure. On the one hand, in the k triangles, the sum of the interior angles is $k \cdot 180°$. On the other hand, in the k triangles, the total of the interior angles whose vertices are 1000 interior points is $1000 \times 360°$ and the total of the interior angles whose vertices are the vertices of the square is $4 \times 90°$. Therefore

$$k \cdot 180° = 1000 \times 360° + 4 \times 90°,$$

and $k = 2002$.

Then, there are $3k$ sides in k triangle. On the other hand, each connecting line segment is a common side of two triangles and every side of the square is one side of some triangle. Therefore $3k = 2l + 4$, and $l = \dfrac{3}{2}k - 2 = 3001$.

Example 2   In a school there are b teachers and c students satisfying the following conditions:

(1) Each teacher teaches exactly k students;

(2) For any two distinct students, there are exactly h teachers teaching them.

Prove that $\dfrac{b}{h} = \dfrac{c(c-1)}{k(k-1)}$. (7th Hong Kong Mathematical Olympiad in 1994)

Proof   If a teacher T_r teaches two students S_i, $S_j (i \neq j)$, then T_r, S_i, S_j form a triple $\{T_r; S_i, S_j\}$ and assume the number of all the triples $\{T_r; S_i, S_j\}$ is M.

On the one hand, for any teacher T_r, he (or she) teaches k students, hence there are $\dbinom{k}{2}$ triples $\{T_r; S_i, S_j\}$ containing T_r, and there are exactly b ways to select T_r, and

$$M = b\binom{k}{2}. \qquad\qquad ①$$

On the other hand, for any two students S_i, $S_j (i \neq j)$, there are exactly h teachers who teach teaching them. Thus there are h triples $\{T_r; S_i, S_j\}$ containing these two students S_i, S_j $(i \neq j)$ and there are exactly $\binom{c}{2}$ ways to select S_i, S_j $(i \neq j)$, and

$$M = h\binom{c}{2}. \qquad \qquad ②$$

From ① and ②, we obtain $b\binom{k}{2} = h\binom{c}{2}$. Thus

$$\frac{b}{h} = \frac{\binom{c}{2}}{\binom{k}{2}} = \frac{c(c-1)}{k(k-1)}.$$

This completes the proof.

Example 3   Suppose that a group of n persons satisfy the following conditions: for any $n - 2$ persons, the number of pairs of mutually acquainted persons are equal, and equals 3^k (k is a positive integer). Find all possible value of n. (China Mathematical Competition in 2000)

Solution   Assume that among all n persons, there are m pairs of mutually acquainted persons. Since n persons may form $\binom{n}{n-2}$ $(n - 2)$-element subsets, and for any $(n - 2)$-element subset, there are 3^k pairs of mutually acquainted persons, hence among all $(n-2)$-element subsets, the total of pairs of mutually acquainted persons equals $3^k\binom{n}{n-2}$.

On the other hand, in the above counting, every pair of mutually acquainted persons belongs to $\binom{n-2}{n-4}$ $(n-2)$-element subsets, hence every pair of mutually acquainted persons is counted repeatedly $\binom{n-2}{n-4}$ times, and

$$m = \frac{3^k \binom{n}{n-2}}{\binom{n-2}{n-4}} = \frac{3^k \binom{n}{2}}{\binom{n-2}{2}} = \frac{n(n-1) \cdot 3^k}{(n-2)(n-3)}. \qquad ①$$

(1) If n is not a multiple of 3, i.e. $(3, n) = 1$, then $(n-3, n) = 1$, $(n-3, 3^k) = 1$ and $(n-2, n-1) = 1$, hence $n-3 \mid n-1$, i.e. $\frac{n-1}{n-3} = 1 + \frac{2}{n-3}$ is a positive integer, and $n-3 \mid 2$. It follows that $n-3 \leqslant 2$, $n \leqslant 5$. But $\binom{n-2}{2} \geqslant 3^k \geqslant 3$, hence $n \geqslant 5$. Therefore $n = 5$.

(2) If n is a multiple 3, then $n-2$ is not a multiple of 3, i.e. $(n-2, 3^k) = 1$ and $(n-2, n-1) = 1$, hence $n-2 \mid n$, i.e. $\frac{n}{n-2} = 1 + \frac{2}{n-2}$ is a positive integer, and $n-2 \mid 2$. It follows that $n-2 \leqslant 2$, $n \leqslant 4$, and it contradicts the result $n \geqslant 5$ which is obtained in (1).

From (1) and (2), we obtain that the unique possible value of n is 5. When $n = 5$, assume that any two of 5 persons is a pair of mutually acquainted persons, thus among any $n - 2 = 3$ persons, there are exactly $\binom{3}{2} = 3^1$ pair of mutually acquainted persons, where $k = 1$ is a positive integer. Hence the conditions of the problem are satisfied. Therefore the required n is 5.

Example 4 Let $2 \leqslant r \leqslant \frac{n}{2}$, and $\Re$ be the family of sets consisting of the r-element subsets of the set $Z_n = \{1, 2, \ldots, n\}$. If the intersection of any two elements (i.e. two r-elements of Z_n) of $\Re$ is a non-empty set, then $|\Re| \leqslant \binom{n-1}{r-1}$ and the equality holds if $\Re = \{A \mid A$ is an r-element set of Z_n and contains a fixed element x of $Z_n\}$. (Erdös-Ko-Rado Theorem)

Proof The number of ways to arrange the n elements of Z_n in a circle is $(n-1)!$. For any circle permutation $\pi_j (j = 1, 2, \ldots, (n-$

1)!), if the elements of set $A_i \in \Re$ is consecutive in π_j, then (A_i, π_j) are paired and let S denote the set of these pairs.

On the one hand, the intersection of any two elements (i. e. two r-element subsets of Z_n) of $\Re$ is a non-empty set, and $2 \leqslant r \leqslant \dfrac{n}{2}$, then for every permutation π_j, there are at most r sets $A_i \in \Re$ such that the elements of set A_i is consecutive in π_j, i. e. there are at most r pairs (A_i, π_j) containing π_j and the number of π_j is $(n-1)!$. Hence

$$|S| \leqslant r \cdot (n-1)!. \qquad ①$$

On the other hand, for every A_i, the number of the circle permutations π_j in which the elements of set A_i is consecutive equals $r!(n-r)!$ and the number of sets A_i is $|\Re|$. Thus

$$|S| = |\Re| \cdot r!(n-r)!. \qquad ②$$

Combining ① with ②, we get

$$|\Re| \cdot r!(n-r)! \leqslant r \cdot (n-1)! \Rightarrow |\Re| \leqslant \frac{r \cdot (n-1)!}{r!(n-r)!} = \binom{n-1}{r-1}.$$

When $\Re = \{A \mid A$ is the r-element set of Z_n and contains a fixed element $x\}$, the number of elements of $\Re$ is $\binom{n-1}{r-1}$, thus the equality $|\Re| = \binom{n-1}{r-1}$ holds.

Remark Further we could prove that the condition of the equality in this problem is a sufficient and necessary condition.

Example 5 Let set $M = \{x_1, x_2, \ldots, x_{4n+3}\}$ and $A_1, A_2, \ldots, A_{4n+3}$ be its $4n+3$ subsets satisfying the following conditions:

(1) Any $n+1$ elements of M belong exactly to a subset $A_i (1 \leqslant i \leqslant 4n+3)$;

(2) $|A_i| \geqslant 2n+1 \ (i = 1, 2, \ldots, 4n+3)$.

Prove that for any two subsets $A_i, A_j (1 \leqslant i < j \leqslant 4n+3)$, they have exactly n common elements.

Proof We construct a $(4n+3) \times (4n+3)$ table such that the number in the ith row and the jth column is

$$a_{ij} = \begin{cases} 1, & \text{if } x_i \in A_j, \\ 0, & \text{if } x_j \notin A_j, \end{cases} \quad i, j = 1, 2, \ldots, 4n+3.$$

Let $r_i = \sum_{j=1}^{4n+3} a_{ij}$, $l_j = \sum_{i=1}^{4n+3} a_{ij}$, then r_i represents that the x_i belongs to the r_i sets of $A_1, A_2, \ldots, A_{4n+3}$ and l_j represents the number of elements of A_j, i. e. $l_j = | A_j |$. From the given condition (2), we obtain

$$\sum_{i=1}^{4n+3} r_i = \sum_{i=1}^{4n+3} \sum_{j=1}^{4n+3} a_{ij} = \sum_{j=1}^{4n+3} \sum_{i=1}^{4n+3} a_{ij} = \sum_{j=1}^{4n+3} | A_j | \geqslant (4n+3)(2n+1).$$

①

If x_k is a common element of two subsets $A_i, A_j (i \neq j)$, then a triple $\{x_k; A_i, A_j\}$ consists of x_k and A_i, A_j.

Let S denote the set of these triples. On the one hand, from the given condition (1), we obtain that for any two subsets $A_i, A_j (i \neq j)$, $| A_i \cap A_j | \leqslant n$, thus there are at most n triples $\{x_k; A_i, A_j\}$ containing the two subsets $A_i, A_j (i \neq j)$, and

$$| S | \leqslant n \binom{4n+3}{2} = n(4n+3)(2n+1).$$

②

On the other hand, every x_k belongs to the r_k subsets of $A_1, A_2, \ldots, A_{4n+3}$, thus there are $\binom{r_k}{2}$ triples $\{x_k; A_i, A_j\}$ containing x_k. Hence

$$| S | = \sum_{k=1}^{4n+3} \binom{r_k}{2} = \frac{1}{2} \left(\sum_{k=1}^{4n+3} r_k^2 - \sum_{k=1}^{4n+3} r_k \right).$$

③

Cobiming ①, ②, and ③ and applying Cauchy's inequality, we obtain

$$n(4n+3)(2n+1) \geqslant \frac{1}{2} \left(\sum_{k=1}^{4n+3} r_k^2 - \sum_{k=1}^{4n+3} r_k \right)$$

$$\geqslant \frac{1}{2} \left(\frac{1}{4n+3} \left(\sum_{k=1}^{4n+3} r_k \right)^2 - \sum_{k=1}^{4n+3} r_k \right)$$

$$= \frac{1}{2(4n+3)} \left(\sum_{k=1}^{4n+3} r_k \right) \left(\sum_{k=1}^{4n+3} r_k - (4n+3) \right)$$

$$\geqslant \frac{1}{2(4n+3)} \cdot (2n+1)(4n+3)[(2n+1)(4n+3) - (4n+3)]$$

$$= n(4n+3)(2n+1).$$

Hence the equality holds in the above inequality, and the equality in ② holds, i.e. for any two subsets A_i, A_j ($1 \leqslant i < j \leqslant 4n+3$), the equality $|A_i \cap A_j| = n$ ($1 \leqslant i < j \leqslant 4n+3$) holds.

Example 6 Let n and k be positive integers and let S be a set of n points in the plane such that:

(1) No three points of S are collinear and

(2) For every point P of S, there are at least k points of S equidistant to P.

Prove that $k < \frac{1}{2} + \sqrt{2n}$. (30th IMO in 1989)

Proof I Let $S = \{P_1, P_2, \ldots, P_n\}$. From the given condition (2), we obtain that for any $P_i \in S$, there is a circle C_i with the center P_i such that there are at least k points of S on the circle C_i ($i = 1, 2, \ldots, n$). Assume that there are exactly r_i points of S on the circle C_i ($i = 1, 2, \ldots, n$), and P_i is the common point of the e_i circles of $C_1, C_2, \ldots, C_n$ ($i = 1, 2, \ldots, n$). Thus

$$\sum_{i=1}^{n} e_i = \sum_{i=1}^{n} r_i \geqslant kn. \qquad \qquad ①$$

If the point P_i of S is the common point of two circles C_t, C_j ($t \neq j$), then a triple $\{P_i; C_t, C_j\}$ consists of P_i and C_t, C_j.

Let M denote the set of these triples. On the one hand, any two circles have at most two intersecting points, then there are at most two triples containing the two circles, thus n circles form at most $2\binom{n}{2}$ triples, and

$$|M| \leqslant 2\binom{n}{2} = n(n-1). \qquad \qquad ②$$

On the other hand, the point P_i is the common point of the e_i circles, thus there are $\binom{e_i}{2}$ triples containing the point $P_i (i = 1, 2, \ldots, n)$, and

$$|M| = \sum_{i=1}^{n} \binom{e_i}{2} = \frac{1}{2} \left(\sum_{i=1}^{n} e_i^2 - \sum_{i=1}^{n} e_i \right). \qquad ③$$

Cobiming ①, ②, and ③ and applying Cauchy's inequality, we obtain

$$n(n - 1) \geqslant \frac{1}{2} \left(\sum_{i=1}^{n} e_i^2 - \sum_{i=1}^{n} e_i \right)$$

$$\geqslant \frac{1}{2} \left(\frac{1}{n} \left(\sum_{i=1}^{n} e_i \right)^2 - \sum_{i=1}^{n} e_i \right).$$

$$= \frac{1}{2n} \left(\sum_{i=1}^{n} e_i \right) \left(\sum_{i=1}^{n} e_i - n \right)$$

$$\geqslant \frac{1}{2n} (kn)(kn - n)$$

$$= \frac{1}{2} nk(k - 1),$$

i.e. $k^2 - k - 2(n - 1) \leqslant 0$. Consequently we arrive at

$$k \leqslant \frac{1 + \sqrt{1 + 8(n - 1)}}{2} < \frac{1 + \sqrt{8n}}{2} = \frac{1}{2} + \sqrt{2n}.$$

This completes the proof.

Proof II   Let $S = \{P_1, P_2, \ldots, P_n\}$. From the given condition (2), we obtain that for any $P_i \in S$, there is a circle C_i with the center P_i such that there are at least k points of S on the circle C_i ($i = 1, 2, \ldots, n$).

The line segment with two end points belonging to S is called a good line segment. Obviously, the number of good line segments equals $\binom{n}{2}$. On the other hand, on every circle, there are at least $\binom{k}{2}$ chords which are good line segments, thus the total of good line segments on n circles are at least $n\binom{k}{2}$. But some common chords are counted repeatedly in the above computation. Since n circles have at most $\binom{n}{2}$ common chords, there are at least $n\binom{k}{2} - \binom{n}{2}$ distinct chords which are good line segments. Summarizing what we have described above, we obtain

$$\binom{n}{2} \geqslant n\binom{k}{2} - \binom{n}{2},$$

i. e.

$$k^2 - k - 2(n-1) \leqslant 0.$$

Therefore we have that

$$k \leqslant \frac{1 + \sqrt{1 + 8(n-1)}}{2} < \frac{1 + \sqrt{8n}}{2} = \frac{1}{2} + \sqrt{2n}.$$

Remark (1) With two proofs above, we see that the condition (1) in this problem is a redundant condition; When we prove the problems using the principle of counting in two ways, since the difference of the counting objects which are selected by us, we often use distinct ways. In this problem, the second proof is more simple than the first proof. But from example 2, example 5 and the latter example, we will see that this way of counting triples is very effective.

Example 7 Prove that

$$\sum_{k=0}^{n} \binom{n}{k} 2^k \binom{n-k}{[(n-k)/2]} = \binom{2n+1}{n}.$$

Proof I On the one hand, the coefficient of x^n in

$$(1+x)^{2n+1} = \sum_{k=0}^{2n+1} \binom{2n+1}{k} x^k$$

is $\binom{2n+1}{n}$.

On the other hand,

$$(1+x)^{2n+1} = (1+x^2+2x)^n(1+x)$$

$$= \sum_{k=0}^{n} \binom{n}{k}(2x)^k(1+x^2)^{n-k} \cdot (1+x) \qquad \textcircled{1}$$

$$= \sum_{k=0}^{n} \binom{n}{k}2^k x^k(1+x^2)^{n-k} + \sum_{k=0}^{n} \binom{n}{k}2^k x^{k+1}(1+x^2)^{n-k}.$$

When $n-k$ is odd, in the expression

$$\binom{n}{k}2^k x^k(1+x^2)^{n-k} = \binom{n}{k}2^k x^k \sum_{i=0}^{n-k}\binom{n-k}{i}x^{2i},$$

does not have a team containing x^n and in the expression

$$\binom{n}{k}2^k x^{k+1}(1+x^2)^{n-k} = \binom{n}{k}2^k x^{k+1} \sum_{i=0}^{n-k}\binom{n-k}{i}x^{2i},$$

the team containing x^n is

$$\binom{n}{k}2^n x^{k+1}\binom{n-k}{(n-k-1)/2}x^{2\cdot(n-k-1)/2} = \binom{n}{k}2^k\binom{n-k}{[(n-k)/2]}x^n.$$

When $n-k$ is even, in the expression

$$\binom{n}{k}2^k x^{k+1}(1+x^2)^{n-k} = \binom{n}{k}2^k x^{k+1} \sum_{i=0}^{n-k}\binom{n-k}{i}x^{2i},$$

there does not have a team containing x^n and in the expression

$$\binom{n}{k}2^k x^k(1+x^2)^{n-k} = \binom{n}{k}2^k x^k \sum_{i=0}^{n-k}\binom{n-k}{i}x^{2i},$$

the team containing x^n is

$$\binom{n}{k}2^n x^k\binom{n-k}{(n-k)/2}x^{2\cdot(n-k)/2} = \binom{n}{k}2^k\binom{n-k}{[(n-k)/2]}x^n.$$

Thus the coefficient of x^n at the right side of $\textcircled{1}$ is $\sum_{k=0}^{n}\binom{n}{k}2^k\binom{n-k}{[(n-k)/2]}$. Therefore we know that

$$\sum_{k=0}^{n} \binom{n}{k} 2^k \binom{n-k}{[(n-k)/2]} = \binom{2n+1}{n}.$$

Proof II We consider a combinatorial mode. There are $2n$ students, n boys and n girls, in a class with their teacher T. Let g_1, $g_2, \ldots, g_n$ denote all the girls, and let $b_1, b_2, \ldots, b_n$ denote all the boys. For $1 \leqslant i \leqslant n$, (g_i, b_i) are paired. The class has n tickets to an exciting soccer game.

We consider the number of ways to find n people to go the game. The obvious answer is $\binom{2n+1}{n}$. On the other hand, we also calculate this number in the following way. For any fixed integer k, $1 \leqslant k \leqslant n$, we find k pairs from n pairs students and give each pair 1 ticket. There are $\binom{n}{k} 2^k$ ways to find k pairs and pick one student from each pair to go to the game. We have $n - k$ tickets left and $n - k$ pairs of students left. We pick $\left[\dfrac{n-k}{2}\right]$ pairs and give each of those pairs 2 tickets. There are $\binom{n-k}{[(n-k)/2]}$ ways to do so. Now we have already assigned $S = k + 2\left[\dfrac{n-k}{2}\right]$ tickets. If $n - k$ is odd, $S = n - 1$, and we assign the last ticket to teacher T, if $n - k$ is even, $S = n$ and we have assigned all the tickets already. It is not difficult to see that as k takes all the values from 1 to n, we obtain all possible ways of assigning the n tickets. Therefore, there are $\sum_{k=0}^{n} \binom{n}{k} 2^k \binom{n-k}{[(n-k)/2]}$ ways to find n people to go the game. Hence

$$\sum_{k=0}^{n} \binom{n}{k} 2^k \binom{n-k}{[(n-k)/2]} = \binom{2n+1}{n}$$

as desired.

Exercise 7

1 There are two students of the first grade who participated in the chess tournament of the second grade. Each player played exactly one game against each of the other players. In each game the winner was awarded 1 point, the loser got 0 point, and each of the two players earned $1/2$ point if the game was a tie. The two students of the first grade earn 8 points in the aggregate, and each student of the second grade has same points as other student of second grade. How many students of the second grade participated in this tournament? How many points does each student of the second grade get?

2 A square is divided into two polygons (containing the triangle) along a straight line, and one of the two polygons is also divided into two polygons along a straight line. Then, one of the three polygons is divided similarly and so on. If we get 47 46-side polygons with dividing the square n times, find the smallest value of n.

3 Prove that among any n people, there exist two people A and B such that among the remaining $n-2$ people, there are at least $\left[\dfrac{n}{2}\right]-1$ people such that each of them and the two people A and B are either mutually acquainted or mutually unacquainted. (When $n=12$, this problem is the examination problem of 48th Moscow Mathematical Olympiad)

4 There are n distinct points in the plane. Prove that the number of point pairs with the unit distance is smaller than $\dfrac{n}{4}+\dfrac{\sqrt{2}}{2}n^{\frac{3}{2}}$.

5 The n elements a_1, a_2, $\ldots$, a_n are paired into n pairs P_1, P_2, $\ldots$, P_n satisfying P_i and P_j have a common element if and only if a_i and a_j are paired. Prove that each of the n elements a_1, a_2, $\ldots$, a_n belong to just two element pairs.

6 Using two methods, prove that for all positive integers, the equality

$$\sum_{i=0}^{[n/2]} 2^{n-2i} \binom{n}{n-2i} \binom{2i}{i} = \binom{2n}{n}$$

holds.

Chapter 8   Recurrence Method

From example 1, example 2, and example 8 in chapter 4, we see that the basic steps to solve combinatorial problems using the recurrence method are:

(1) To find the initial values of the sequence using the simple enumeration;

(2) To establish a recurrence relation;

(3) To find the solution by the recurrence relation.

When solving some problem is difficult by the given recurrence relation, we often need to establish a new recurrence relation. In some case, establishing a new recurrence relation is also difficult, we may enumerate several especial cases and find the required recurrence relation by induction.

Example 1 Let the permutation of positive integers $1, 2, \ldots, n$ satisfy the following condition: each number is greater than or less than all numbers in front of them. How many such permutations are there? (21st Canadian Mathematical Olympiad in 1989)

Solution Let a_n denote the required number of permutations. Obviously $a_1 = 1$, $a_2 = 2$. For $n > 2$, we consider the permutation in which the ith number is n. As the permutation of $n - i$ numbers behind n are determined uniquely, i. e. $n - i$, $n - i - 1$, $\ldots$, 2, 1 and the number of the permutations of the $i - 1$ numbers in front of n is a_{i-1} ($i = 1, 2, \ldots, n$ and $a_0 = 1$). Thus

$$a_n = 1 + a_1 + a_2 + \cdots + a_{n-1}, \qquad \text{①}$$

and

$$a_{n-1} = 1 + a_1 + a_2 + \cdots + a_{n-2}. \qquad ②$$

With ① − ②, we obtain $a_n - a_{n-1} = a_{n-1}$, i. e. $a_n = 2a_{n-1}$.
Therefore $a_n = a_1 \cdot 2^{n-1} = 2^{n-1}$.

Example 2 There are n boys and n girls in the city A and each girl knows all the boys, and there are n girls: g_1, g_2, $\ldots$, g_n and $2n - 1$ boys: b_1, b_2, $\ldots$, b_{2n-1} in the city B such that the girl g_i just knows each of the boys b_1, b_2, $\ldots$, b_{2i-1} ($i = 1, 2, \ldots, n$). For any $r \in \{1, 2, \ldots, n\}$, we select r girls and r boys in the city A as well as in the city B such that they are matched into r pairs of partners in a dance satisfying the following condition: the partner of the girl in each pair is known by herself. Let $A(r)$ and $B(r)$ denote the number of the distinct ways to select r girls and r boys in the city A and in the city B respectively. Prove that $A(r) = B(r)$. (The Problem Prepared for 38th IMO)

Proof For convenience, let $A_n(r)$ and $B_n(r)$ denote $A(r)$ and $B(r)$ respectively. Since each girl knows all the boys in the city A, for $n \geqslant 1$,

$$A_n(r) = \binom{n}{r}^2 \cdot r! = \frac{(n!)^2}{[(n-r)!]^2 \cdot r!}. \qquad ①$$

Now we establish the recurrence relation of $B_n(r)$ for $n \geqslant 3$, $2 \leqslant r \leqslant n$.

Case I : When g_n is one of the r girls who are selected, then the number of the distinct ways to choose the remaining $r - 1$ pairs is $B_{n-1}(r-1)$. Since the partner of g_n can be any one of the remaining $(2n-1) - (r-1) = 2n - r$ boys, in this case, by the multiplication principle, the number of distinct ways is $(2n-r)B_{n-1}(r-1)$.

Case II : When g_n is not one of the r girls who are selected, then $r < n$ and the r girls are selected from g_1, g_2, $\ldots$, g_{n-1} and their partners are from b_1, b_2, $\ldots$, b_{2n-3}. In this case, the number of the distinct ways is $B_{n-1}(r)$.

Summarizing what we have described above, we conclude that for $n \geqslant 3$,

$$B_n(r) = B_{n-1}(r) + (2n - r)B_{n-1}(r-1) \Big\rbrace$$
$$B_n(n) = nB_{n-1}(n-1) \Big\rbrace \qquad ②$$

Where $r = 1, 2, \ldots, n-1$.

Obviously $A_n(1) = n^2 = B_n(1)$ and $A_2(2) = 2 = B_2(2)$. From ①, we know

$$A_{n-1}(r) = \frac{[(n-1)!]^2}{[(n-1-r)!]^2 r!},$$

and

$$A_{n-1}(r-1) = \frac{[(n-1)!]^2}{[(n-r)!]^2 (r-1)!}.$$

From the above equalities, we have

$$A_n(r) = A_{n-1}(r) + (2n-r)A_{n-1}(r-1) \Big\rbrace$$
$$A_n(n) = nA_{n-1}(n-1) \Big\rbrace \qquad ③$$

From ② and ③, we could deduce that the two sequences $A_n(r)$ and $B_n(r)$ have the same initial values and the same recurrence relation, hence for any $n \in \mathbf{N}_+$ and $r = 1, 2, \ldots, n$, $A_n(r) = B_n(r)$.

This completes the proof.

Example 3   A circular disk is divided into n ($n \geqslant 2$) sectors S_1, $S_2, \ldots, S_n$ and each sector is colored with one of m colors such that the colors of any two adjacent sectors are different. How many distinct coloring ways are there?

Solution   Let the number of distinct coloring ways is $a_n(m)$.

(1) Find the initial values. When $n = 2$, then S_1 could be colored with one of m colors, and S_2 could be colored with one of the remaining colors, by the multiplication principle, we know

$$a_2(m) = m(m-1).$$

(2) Find the recurrence relation. Since S_1 could be colored with one of m colors, and S_2 could be colored with one of $m - 1$ colors, $\ldots$, S_{n-1} could be colored with one of $m - 1$ colors, and S_n could be colored with one of $m - 1$ colors (the colors of S_n and S_1 may

be the same or distinct.), there are $m(m-1)^{n-1}$ distinct ways.

Case I : The colors of S_n and S_1 are distinct. In this case, there are $a_n(m)$ distinct coloring ways.

Case II : The colors of S_n and S_1 are the same. In this case, S_n and S_1 the sectors S_n and S_1 are incorporated into a bigger sector, and we know that the colors of S_1 and S_2 are distinct, thus there are $a_{n-1}(m)$ distinct coloring ways

By the addition principle, we obtain

$$a_n(m) + a_{n-1}(m) = m(m-1)^{n-1}(n \geqslant 2). \qquad ①$$

(3) Find $a_n(m)$. Let $b_n(m) = \dfrac{a_n(m)}{(m-1)^n}$. From ①, we know

$$b_n(m) + \frac{1}{m-1}b_{n-1}(m) = \frac{m}{m-1},$$

i. e.

$$b_n(m) - 1 = -\frac{1}{m-1}(b_{n-1}(m) - 1).$$

$\left(\text{Where 1 is the root of the equation } x + \dfrac{1}{m-1}x = \dfrac{m}{m-1}.\right)$

Thus

$$b_n(m) - 1 = (b_2(m) - 1)\left(-\frac{1}{m-1}\right)^{n-2}$$

$$= \left[\frac{a_2(m)}{(m-1)^2} - 1\right]\left(-\frac{1}{m-1}\right)^{n-2}$$

$$= \left[\frac{m(m-1)}{(m-1)^2} - 1\right]\left(-\frac{1}{m-1}\right)^{n-2}$$

$$= (-1)^n \frac{1}{(m-1)^{n-1}},$$

and

$$a_n(m) = (m-1)^n + (-1)^n(m-1),$$

i. e. there are $(m-1)^n + (-1)^n(m-1)$ distinct coloring ways.

Example 4 How many ways are there to color the $n+1$ vertices of an n-pyramid $S - A_1A_2 \cdots A_n$ with $n+1$ colors such that each of the vertices is colored with a color and two vertices with a common edge must be colored with different colors? (Especially, when $n = 4$, this problem is the problem of China Mathematical Competition in 1995)

Solution   There are $\binom{n+1}{1}$ ways to select one from the $n + 1$ colors to paint the vertex S. Since each of the n vertices A_1, A_2, ..., A_n in the base must be colored with one of the remaining n colors such that each vertex is colored with a color and two vertices with a common edge must be colored⁣ with different colors, then from example 3, we obtain that the number of the coloring ways is $a_n(n)$. By the multiplication principle, the required number of distinct ways is

$$\binom{n+1}{1}a_n(n) = (n+1)[(n-1)^n + (-1)^n(n-1)].$$

(Especially, when $n = 4$, the required number of the distinct ways is $5[3^4 + 3] = 420$.)

Example 5   A circle with the perimeter 24 is divided into 24 equal arcs. How many ways are there to choose 8 points from 24 points of division such that the length of the arc between any two points of division is not equal to 3 or 8? (CMO in 2001)

Solution   Let 24 points of division be denoted by 1, 2, ..., 24. The 24 numbers are arranged into a 3×8 table as follows:

1	4	7	10	13	16	19	22
9	12	15	18	21	24	3	6
17	20	23	2	5	8	11	15

In the above table, the two adjacent numbers in each row correspond to two points (the first grid and the 8th grid in the same row are regarded as adjacent) and the length of the arc between them equals 3, and the two adjacent numbers in each column correspond to two points (the first grid and the third grid in the same column are regarded as adjacent) and the length of the arc between them equals 8.

Thus we choose at most one number in each column. But we must choose 8 numbers, hence we choose exactly one number in each column and the two numbers which are chosen in the two adjacent columns are not in the same row (the first column and the 8th column are regarded as adjacent).

If we regard each column as a sector, and the first row, the second row and the third row of each column are regarded as 3 colors, then this problem is equivalent to the case of $n = 8$, $m = 3$ in the example 3.

Hence the required number of ways to choose 8 points of division is

$$a_8(3) = 2^8 + (-1)^8 \cdot 2 = 258.$$

Example 6 Let a_1, a_2, $\ldots$, a_n be any permutation of 1, 2, $\ldots$, n and let $f(n)$ denote the number of the permutations satisfying the following conditions:

(1) $a_1 = 1$;

(2) $|a_i - a_{i+1}| \leqslant 2$, $i = 1, 2, \ldots, n - 1$.

It is possible that $f(2004)$ is divisible by 3?

Solution The permutation a_1, a_2, $\ldots$, a_n is called a good permutation satisfying the conditions (1) and (2). Since $a_1 = 1$, then $a_2 = 2$ or 3.

(I) If $a_2 = 2$, then $a_2 - 1$, $a_3 - 1$, $\ldots$, $a_n - 1$ is also a good permutation, and the number of these permutations is $f(n - 1)$.

(II) If $a_2 = 3$, $a_3 = 2$, then $a_4 = 4$, and $a_4 - 3$, $a_5 - 3$, $\ldots$, $a_n - 3$ is also a good permutation, and the number of these permutations is $f(n - 3)$.

(III) If $a_2 = 3$, $a_3 \geqslant 4$, let a_{k+1} be the first even in this permutation (from the left to right), then a_1, a_2, $\ldots$ a_k must be 1, 3, 5, $\ldots$, $2k - 1$ and $a_{k+1} = 2k$ or $2k - 2$. Hence a_k and a_{k+1} are two adjacent positive integers. From the condition (2), we obtain that all the numbers behind a_{k+1} are larger or less than a_{k+1}. Since 2 is behind a_{k+1}, a_{k+2}, a_{k+3}, $\ldots$, a_n are all less then a_{k+1}. Therefore this permutation is unique, i. e. all positive odds less than or equal to n are arranged in an increasing order and then all positive even less than or equal to n are arranged in a decreasing order.

Summarizing what we have described above, we obtain the following recurrence relation:

$$f(n) = f(n-1) + f(n-3) + 1 \ (n \geqslant 4).$$

Using the method of enumeration, we obtain $f(1) = 1$, $f(2) = 1$, $f(3) = 2$, and with the above recurrence relation, we may count the remainder with modulo 3 of this sequence, in the order of

$$1, 1, 2, 1, 0, 0, 2, 0, 1, 1, 2, 1, 0, 0, 2, 0, \cdots$$

Thus the period of this sequence is 8 (this conclusion can be proved easily by the mathematical induction), and $2004 = 125 \times 8 + 4$, and $f(2004) \equiv 1 \pmod{3}$, i. e. $f(2004)$ is not divisible by 3.

Example 7   Are there infinitely many pairs of positive integers (a, b) such that $a \mid b^2 + 1$ and $b \mid a^2 + 1$?

Analysis   We enumerate several pairs of the positive integers satisfying the condition as follows:

a	1	2	5	13	34	$\cdots$
$a^2 + 1$	2	5	26	170	1157	$\cdots$
b	2	5	13	34	89	$\cdots$
$b^2 + 1$	5	26	170	1157	7922	$\cdots$

From the above table, we see that the values of a, b are exactly two adjacent terms of the following sequence: $a_1 = 1$, $a_2 = 2$, $a_{n+2} = 3a_{n+1} - a_n (n \in \mathbf{N}_+)$ (This recurrence relation could be obtained by the following method: let $a_{n+2} = pa_{n+1} + qa_n$. (Substituting $a_1 = 1$, $a_2 = 2$, $a_3 = 5$, and $a_4 = 13$ to the above expression, we obtain $p = 3$, $q = -1$, thus $a_{n+2} = 3a_{n+1} - a_n$.) and the equality $a_{n+1}^2 + 1 = a_n a_{n+2}$ holds.

Solution I   Firstly, we prove the following lemma.

Lemma 1   Let $a_1 = 1$, $a_2 = 2$, and $a_{n+2} = 3a_{n+1} - a_n (n \in \mathbf{N}_+)$, then $a_{n+1}^2 + 1 = a_n a_{n+2}$.

Proof of lemma 1 When $n = 1$,

$$a_3 = 3a_2 - a_1 = 3 \times 2 - 1 = 5,$$
$$a_2^2 + 1 = 2^2 + 1 = 1 \times 5 = a_1 a_3.$$

Suppose that when $n = k$, $a_{k+1}^2 + 1 = a_k a_{k+2}$, and when $n = k+1$, we know

$$\begin{aligned}
a_{k+2}^2 + 1 &= a_{k+2}(3a_{k+1} - a_k) + 1 \\
&= 3a_{k+2}a_{k+1} - (a_k a_{k+2} - 1) \\
&= 3a_{k+2}a_{k+1} - a_{k+1}^2 \\
&= a_{k+1}(3a_{k+2} - a_{k+1}) = a_{k+2}a_{k+3}.
\end{aligned}$$

This completes the proof.

Next, let $a = a_{n+1}$, $b = a_{n+2} \ (n \in \mathbf{N}_+)$, then there are infinitely many pairs of positive integers (a, b) and

$$a^2 + 1 = a_{n+1}^2 + 1 = a_n a_{n+2} = a_n b$$

is divisible by b and

$$b^2 + 1 = a_{n+2}^2 + 1 = a_{n+1}a_{n+3} = a a_{n+3}$$

is divisible by a.

Remark In this problem, the values of a, b are exactly the adjoining odd terms of the following Fibonacci sequence:

$$1, 1, 2, 3, 5, 8, 13, 21, 34, 55, 89, \ldots.$$

Therefore we could yield another proof of this problem.

Solution II Firstly, we prove the following lemma.

Lemma 2 Let $f_1 = 1$, $f_2 = 1$, and $f_{n+2} = f_{n+1} + f_n \ (n \in \mathbf{N}_+)$, then $f_{2n+1}^2 + 1 = f_{2n-1}f_{2n+3}$.

Proof of lemma 2 Since

$$\begin{aligned}
f_{2n+1} &= f_{2n} + f_{2n-1} \\
&= f_{2n-1} + f_{2n-2} + f_{2n-1} \\
&= 2f_{2n-1} + (f_{2n-1} - f_{2n-3}) \\
&= 3f_{2n-1} - f_{2n-3},
\end{aligned}$$

let $a_n = f_{2n-1}$, we can deduce lemma 2 from lemma 1.

Next let $a = f_{2n+1}$, and $b = f_{2n+3}$, then there exist infinitely many pairs of positive integers (a, b) such that $a \mid b^2 + 1$ and $b \mid a^2 + 1$.

Example 8　Are there infinitely many triangles such that the lengths of three sides are relatively prime positive integers and the areas are a perfect squares?

Solution　Suppose that there exists a such triangle with the lengths x, y, z of three sides (x, y, z are three relatively prime positive integers). Denote $p = \frac{1}{2}(x + y + z)$, then the area of the triangle is

$$S = \sqrt{p(p-x)(p-y)(p-z)}. \qquad ①$$

For convenience, let $p - z = 1$, i. e, $z = x + y - 2$, $p = x + y - 1$, $p - x = y - 1$. Substituting these to ①, we obtain

$$S = \sqrt{(x-1)(y-1)(x+y-1)}. \qquad ②$$

Let $y - 1 = (a-1)x (a \in \mathbf{N}_+)$, then $x + y - 1 = ax$. Substituting these into ②, we obtain

$$S = \sqrt{(x-1)(a-1)x \cdot ax} = x\sqrt{a(a-1)(x-1)}. \qquad ③$$

Since S must be a positive integer, we assume that $x - 1 = a(a-1)$. From ③, we obtain

$$S = x(x-1) = a(a-1)[a(a-1)+1].$$

But such S is not always a perfect square, hence we assume that $x - 1 = 4a(a-1)$, i. e. $x = (2a-1)^2$, and

$$S = a(2a-2)(2a-1)^2.$$

Therefore we need to seek infinitely many positive integers a such that a and $2a - 2$ are perfect squares. Obviously $a = 9$ is a solution. Generally, we could take $a = a_n$ where a_n satisfies the following recurrence relation: $a_1 = 9$, $2a_{n+1} - 2 = 4a_n(2a_n - 2)$, i. e. $a_{n+1} = (2a_n - 1)^2$. Thus for any $n \in \mathbf{N}_+$, a_n and $2a_n - 2$ are perfect squares. In fact, $a_1 = 9$ and $2a_1 - 2 = 16$ are the perfect squares. Assume that a_k

and $2a_k - 2$ are the perfect squares, then $a_{k+1} = (2a_k - 1)^2$ and $2a_{k+1} - 2 = 4a_k(2a_k - 2)$ are also the perfect squares. Hence for any $n \in \mathbf{N}_+$, a_n and $2a_n - 2$ are the perfect squares.

We come back to the original problem. Assume that the lengths of three sides of the triangle are

$$x = (2a_n - 1)^2,$$
$$y = (a_n - 1)x + 1 = (a_n - 1)(2a_n - 1)^2 + 1,$$
$$z = x + y - 2 = a_n(2a_n - 1)^2 - 1,$$

then $(x, y) = 1$, $(y, z) = 1$, $(z, x) = 1$, i.e. x, y, z are relatively prime and

$$z - y = (2a_n - 1)^2 - 2 > 0,$$
$$y - x = (a_n - 2)(2a_n - 1)^2 + 1 > 0.$$

That is $z > y > x$ and

$$x + y = a_n(2a_n - 1)^2 + 1 > a_n(2a_n - 1)^2 - 1 = z.$$

Therefore there exists a triangle in which the lengths of three sides are x, y, z, and the area of this triangle is $S = a_n(2a_n - 2)(2a_n - 1)^2$ which is a perfect square. This completes the proof.

Remark From example 7 and example 8, we could find that in the combinatorial number theory and the combinatorial geometry, we often need to prove that there exist infinitely many sequences of numbers or figures satisfying some conditions (these conditions often relate to the positive integers). To seek the solutions of these problems, we could start with the especial cases or the simple cases, and apply the methods of exploration and induction to find the common rule. Afterward we could complete this proof by constructing the corresponding recurrence sequence.

Exercise 8

1 In an athletic meet, m medals were awarded during n days.

On the first day one medal and a seventh of the remaining were awarded. On the second day two medals and a seventh of the remaining were awarded, and so on, and on the n th day, the last n medals were awarded. How many days did the competition run over and how many medals had been awarded? (9th IMO)

2   A plane is divided into two parts by an ellipse. A plane is at most divided into six parts by two ellipses. How many parts at most are there if the plane is divided by 10 ellipses?

3   Let $n \geqslant 2$ be a positive integer. How many permutations $\{a_1, a_2, \ldots, a_n\}$ of $\{1, 2, \ldots, n\}$ are there such that $a_i - a_{i+1}$ is divisible by i ($i = 1, 2, \ldots, n-1$) and find all these permutations.

4   An n-term sequence in which each term is either 0 or 1 is called a binary sequence of length n. Let A be a binary sequence with the finite length, and $f(A)$ represent a transformation such that each 1 in A becomes 0, 1 and each 0 in A becomes 1, 0. As an example,

$$f((1, 0, 0, 1)) = (0, 1, 1, 0, 1, 0, 0, 1).$$

Find the number of pairs containing two consecutive 0 in $f^{(n)}((1))$, where $f^{(1)}(A) = f(A)$, $f^{(k+1)}(A) = f^{(k)}(A)$, $k = 1, 2, \ldots$.

5   An increasing sequence of numbers is called an alternative sequence if the first term is odd, the second term is even, the third term is odd, and the fourth term is even and so on. The empty set is also regarded as a alternative sequence. Let $A(n)$ denote the number of the alternative sequences whose terms are taken from the set $\{1, 2, \ldots, n\}$. Obviously, $A(1) = 2$, and $A(2) = 3$. Find $A(20)$.

6   How many positive numbers less than 10^4 are there such that $2^n - n^2$ is divisible by 7? (6th Moscow Mathematical Olympiad)

7   Are there infinitely many pairs (a, b) of the positive integers such that $a^2 + b^2 + 1$ is divisible by ab?

8   Are there infinitely many positive integers triple (a, b, c) such that a, b, c construct an arithmetic sequence $(a < b < c)$ and $ab + 1, bc + 1, ca + 1$ are all the perfect squares.

9   Are there infinitely many triangles $\triangle ABC$ such that lengths of three sides AB, BC, CA construct an arithmetic sequence consisting of the positive integers ($AB < BC < CA$) and the altitude on the side BC and area of $\triangle ABC$ are two positive integers.

Chapter 9   Coloring Method and Evaluation Method

9.1   Coloring Method

According to the characters of the investigative problems, each investigative object is colored with one of several colors, thus many combinatorial objects (the points, the lines, the regions, ...) and the combinatorial constructions (the angles with the same color sides, the angles with the distinct color sides, the triangles with three same color sides, the triangles with not all the same color sides, the pairs of the same color points, the pairs of distinct color points, ...) are formed. Afterward, for the numbers and properties of these objects and constructions, we make an analysis and comparison in detail such that the answer of the investigative problem could be found easily. It is the coloring method.

Example 1   When two adjacent unit squares are moved from a row of the 2×3 rectangle, the obtained shape is called an "L-shape". There is an 8×8 chessboard. Is it possible to obtain a perfect cover of this chessboard using 7 2×2 squares and 9 "L-shapes"? (Reasoning is required.)

Solution   Let each unit square in the odd rows of the 8×8 chessboard be colored white and each unit square in the even rows of the 8×8 chessboard be colored black. Thus there are 32 white unit squares and 32 black unit squares in this chessboard.

If we can obtain a perfect cover of this chessboard using 7 squares (2×2) and 9 "L-shapes", then each 2×2 squares covers 2 white unit squares and 2 black unit squares, and each "L-shape" covers 1 white unit square and 3 black unit squares or 3 white unit squares and 1 black unit square. Assume that there are x "L-shapes" who cover 1 white unit square and 3 black unit squares, and the remaining $9 - x$ "L-shapes" who cover 3 white unit squares and 1 black unit square. Thus the number of the black unit squares covered is $7 \times 2 + 3x + (9 - x) =$ 32, and $x = \dfrac{9}{2}$. But $\dfrac{9}{2}$ is not an integer, so it leads to a contradiction. Therefore there is no perfect cover satisfying the conditions of the problem.

Example 2 Let a 23×23 ground is covered with 1×1, 2×2, 3×3 ceramic tiles. How many 1×1 ceramic tiles are used at least? (Assume that each ceramic tile cannot be divided into little tiles.)

Solution Let the 23×23 square ground be divided into 23^2 unit squares (1×1), and all unit squares of the 1st, 4th, 7th, ..., 19th, 22th column are colored black, and other unit squares are colored white, then each 2×2 ceramic tile covers 2 white unit squares and 2 black unit squares or 4 white unite squares, and each 3×3 ceramic tile covers 3 black unit squares and 6 white unite squares. Assume there is no a 1×1 ceramic to use. No matter how many 2×2 and 3×3 ceramic tiles are used, the number of white unit squares which are covered in the 23×23 ground is even. But the number of white unit squares in the 23×23 ground equals 15×23 which is odd. It is a contradiction. Therefore the number of the 1×1 ceramic tiles is used at least one.

On the other hand, as in figure 9.1, we know that the 12×11 rectangle ground may be covered with 2×2, 3×3 ceramic tiles and as in figure 9.2, the 23×23 ground be covered with only one 1×1 ceramic tile and four 12×13 rectangles which consist of the 6 ceramic tiles (2×2) and 12 ceramic tiles (3×3).

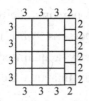

Figure 9. 1

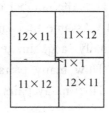

Figure 9. 2

Summarizing what we have described above, the required smallest value of the number of the 1×1 ceramic tiles is 1.

Example 3 Suppose there are 30 persons attending a parliament, and any two persons are friends or political opponents, and each person has 6 political opponents exactly. Three persons form a triple and a triple is called a good triple if any two persons of the triple are friends or opponents. Find the number of the good triples. (24th USSR Mathematical Olympiad)

Solution Let us represent 30 persons with 30 points in space, no four points of which are coplanar. If two persons are friends then the two corresponding points are connected by a red line segment, otherwise, by a blue line segment. Thus the number of the good triple equals the number of the monochromatic triangles (i. e. the three sides of the triangle are colored with the same color). If the two sides of an angle have distinct colors, then the angle is called a distinct color angle. Thus there is not a distinct color angle in each monochromatic triangle and there are exactly two distinct color angles in each triangle which is not a monochromatic triangle. Since the number of the blue line segments meeting each point is 6, and the number of the red line segments meeting each point is 23, the number of the distinct color angle of every point is $23 \times 6 = 138$. The total of the distinct color angles in figure is $138 \times 30 = 4140$. Therefore the number of the triangles which is not the monochromatic triangles is equal to $\frac{1}{2} \times 4140 = 2070$ and the number of the monochromatic triangles is equal to $\binom{30}{3} - 2070 = 1990$. Thus the number of the good triples is 1990.

Example 4 Assume that among eight persons, neither 3 persons who are mutually acquainted nor 4 persons who are mutually unacquainted. How many pairs of mutually acquainted persons are there at least? How many pairs of mutually unacquainted persons are there at least?

Solution Let us represent 8 persons with 8 points in the plane. If two persons are mutually acquainted then the two corresponding points are connected by a red line segment, otherwise by a blue line segment. Thus we obtain a 2-colored complete graph K_8 and from the given conditions, we conclude that there is no a red triangle and a blue complete graph K_4 in this 2-colored complete graph K_8.

If there are at least 6 blue sides meeting some point, by Ramsey's theorem, in the 2-colored complete graph K_6 whose 6 vertices are ends of these 6 blue sides, there exists a red or blue triangle. But the blue triangle leads to a blue complete graph K_4, and it contradicts the given conditions. Hence in this complete graph K_8, the number of the red sides meeting every point is at least 2, and there are at least $\frac{1}{2} \times 2 \times 8 = 8$ red sides in this graph K_8.

If there are at least 4 red sides meeting some point, then the complete graph K_4 whose 4 vertices are ends of the 4 red sides is blue or has at least a red side. But having a red side leads to a red triangle and it contradicts the given conditions. Hence there are at most 3 red sides meeting every point of K_8, and there are at more $\frac{1}{2} \times 3 \times 8 = 12$ red sides.

(1) If there are exactly 8 red sides in the graph K_8, then there are exactly 2 red sides meeting every point. Hence the 8 red sides construct one or two cycles (the closed broken line). If there are two cycles, and there are exactly two red sides meeting every point, and the two cycles have no common vertex and since there is no red triangle in this graph. Hence each cycle is a red 4-side polygon. If two red 4-side polygons $A_1A_2A_3A_4$ and $A_5A_6A_7A_8$ consist of the 8 red sides, then there is blue complete graph K_4 with the vertices A_1, A_3, A_5, A_7. If the red 8-side polygon $A_1A_2\cdots A_8$ consists of the 8 red sides, then there is also a blue complete graph K_4 with the vertices A_1, A_3, A_5, A_7. They all contradict the given conditions. Hence there are at least 9 red sides in this graph K_8.

(2) Assume there are exactly 9 red sides in the graph K_8. Since

there are at least 2 and at most 3 red sides meeting every point and there is not a red triangle in this K_8, K_8 could be one of the figure 9. 3 (a) or the figure 9. 3(b). (The solid line segments represent the red sides, but we do not draw the blue sides.) In the figure 9. 3(a), the dashed line segment represents that the line segment A_2A_7 may is replaced by the line segment A_2A_6. It is easy to find that in the figure 9. 3(a) and 9. 3(b), there is the blue complete graph K_4 with the vertices A_1, A_3, A_5, A_7. It contradicts the given conditions.

Therefore there are at least 10 red colored sides in this K_8.

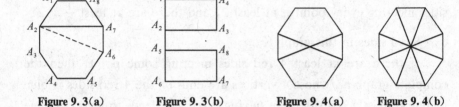

Figure 9. 3(a)   Figure 9. 3(b)   Figure 9. 4(a)   Figure 9. 4(b)

On the other hand, in the figure 9. 4(a) and 9. 4(b), there is neither a red triangle nor a blue complete graph K_4.

Summarizing what we have described above, we obtain that in the 2-colored complete graph K_8, there are at least 10 red sides and at most 12 red sides, i. e. among 8 persons, there are at least 10 pairs of the mutually acquainted persons and at most 12 pairs of the mutually acquainted persons. Hence there are at least $\binom{8}{2} - 12 = 16$ pairs of the mutually unacquainted persons.

9. 2 Evaluation Method

According to the characters of the investigative problems, the investigative objects are evaluated with distinct numbers respectively, and we analyze, compute and compare these numbers in detail such that the answer of the investigative problem could be found easily. It is the evaluation method.

Example 5 When a unit square is moved from a 2×2 square, the obtained shape is called a "Γ-shape". Use some "Γ-shapes" to cover a 5×7 rectangle, and each "Γ-shape" covers exactly 3 unit squares. Assume that the overlaps are allowed and without gaps and each "Γ-shape" does not exceed the sides of the 5×7 rectangle. Is it possible that each unit square of the 5×7 rectangle is covered by the same number of "Γ-shapes". Justify your answer. (22^{sd} Russia Mathematical Olympiad)

Solution Suppose that we write -2 or 1 in each unit square of the 5×7 rectangle as in figure 9. 5, then the sum of any three numbers in the square which are covered by a "Γ-shape" is an nonnegative number. Hence no matter how many times the numbers are covered by the "Γ-shapes", the sum of all numbers which are covered by these "Γ-shapes" (if one number is covered n times then it is computed n times in this sum) is also a nonnegative number. On the other hand, the sum of the numbers in the figure 9. 5 rectangle is $12 \times (-2) + 23 \times 1 = -1$. If each unit square is covered k times, then the total of all numbers which are covered by these "Γ-shapes" is equal to $-k$. It is a contradiction. Therefore it is impossible that each unit square of the 5×7 rectangle is covered by the same number of "Γ-shapes".

-2	1	-2	1	-2	1	-2
1	1	1	1	1	1	1
-2	1	-2	1	-2	1	-2
1	1	1	1	1	1	1
-2	1	-2	1	-2	1	-2

Figure 9. 5

Example 6 A regular hexagon is divided into 24 equilateral little triangles by lines parallel to its sides. 19 different real numbers are assigned to the 19 vertices of 24 little triangles. Show that at least 7 of 24 triangles have the property that numbers assigned to its vertices may construct a decreasing sequence in clockwise. (19th USSR Mathematical Olympiad)

Proof The 24 triangles are divided into two parts: in the part (1), three real numbers assigned to the vertices of the triangles may construct a decreasing sequence in clockwise and the remaining triangles belong to the part (2). Assume that the number of the triangles in the part (1) is N, then the number of the triangles in the part (2) is $24 - N$. For every side of the little triangles, we draw a arrow from the vertex corresponding to a big number to the vertex corresponding a smaller number (implies decreasing!) and we write -1 on the left side of this arrow and 1 on the right side of this arrow. (The right side indicates that the direction of this arrow is clockwise and the left side indicates that the direction of this arrow is counter-clockwise!). Thus the sum of three numbers in every triangle in the part (1) is $1 + 1 + (-1) = 1$ and the sum in the part (2) is $1 + (-1) + (-1) = -1$ as in figure 9.6. Hence the total of the numbers in the 24 triangles is

$$N + (24 - N) \times (-1) = 2N - 24.$$

Figure 9.6

On the other hand, the sum of the two numbers in the two sides of every short line segments in the regular hexagon is $(-1) + 1 = 0$, and among 12 short line segments on the sides of the regular hexagon which are the sides of 12 little regular triangles respectively, there is at least one short line segment such that the number in the interior side of the regular hexagon is 1. Hence the total of the numbers on the sides of the regular hexagon is at least $11 \times (-1) + 1 = -10$.

Summarizing what described above, we gather that $2N - 24 \geqslant -10$, i.e. $N \geqslant 7$. This completes the proof.

Remark Example 6 may be generalized to the following conclusion: let a polygon be divided perfectly into little triangles such that each vertex of every triangle is not an interior point of sides of another triangle. (as figure 9.7) Suppose that the total number of little triangles is n and the number of the triangles with a side on the sides of this

Figure 9.7

polygon is m and $n - m$ is even. We write any distinct real numbers at these vertices. Prove that among n triangles, there are at least $\frac{1}{2}(n - m) + 1$ triangles such that three real numbers written at the vertices of every triangle could be arranged in the decreasing sequence in clockwise.

Example 7 There are 20 boys and 20 girls, standing face to face in two concentric circles and there are exactly 20 persons in each circle with no stipulation on the number of the boys and girls. Let each person in outer circle and the opposite one person in inner circle be paired. If two persons in the same pair are a boy and a girl, then they form a pair of partners. Prove that it is possible to rotate persons in inner circle to proper position so that the number of pairs of partners is at least 10.

Proof Let 20 persons in inner (outer) circle be labeled clockwise with the numbers $a_i (b_i)$ ($i = 1, 2, \ldots, 20$) in clockwise order, such that for the boys, $a_i (b_i)$ is 1, and for the girls, $a_i (b_i)$ is -1. To rotate persons in inner circle such that the person with the label a_1 in inner circle and the person with label b_i in outer circle are opposite and let

$$S_i = \sum_{k=1}^{20} a_k b_{i+k-1} (i = 1, 2, \ldots, 20; \; b_{j+20} = b_j).$$

Without loss of generality, assume that the number of girls is greater than or equal to the number of boys in inner circle, thus the number of boys is greater than or equal to the number of girls in outer circle, which implies

$$a_1 + a_2 + \cdots + a_{20} \leqslant 0,$$
$$b_1 + b_2 + \cdots + b_{20} \geqslant 0.$$

Thus

$$S_1 + S_2 + \cdots + S_{20} = (a_1 + a_2 + \cdots + a_{20})(b_1 + b_2 + \cdots + b_{20}) \leqslant 0.$$

By the mean value principle, there exists an i_0 ($1 \leqslant i_0 \leqslant 20$) such that $S_{i_0} \leqslant 0$. It means that in the expression $a_1 b_{i_0} + a_2 b_{i_0+1} + \cdots +$

$a_{20}b_{i_0+19}(b_{j+20} = b_j)$, the number of terms which equal -1 are not less than the number of terms which equal 1, so there are at least 10 pairs of partners.

Example 8   All the grids of an $m \times n$ chessboard ($m \geqslant 3$, $n \geqslant 3$) are colored with red or blue. Two adjacent grids (with a common side) are called a good couple if they are of different colors. Suppose that there are S good couples, and explain how to determine whether S is odd or even. Does it depend on certain specific color grids? (Reasoning is required.) (4th China Western Mathematical Olympiad in 2004)

Solution   Classify all grids into three parts: the grids at the four corners, the grids along the border-lines (not including four corners), and the other grids. Fill all red grids with label number 1, all blue grids with label number -1. Denote the label numbers filled in the grids in the first part by a, b, c and d, in the second part by x_1, $x_2, \ldots, x_{2m+2n-8}$, and in the third part by y_1, $y_2, \ldots, y_{(m-2)(n-2)}$. For any two adjacent grids we write a label number which is the product of two label numbers of the two grids on their common edge. Let H be the product of all the label numbers on the common edges. There are 2 adjacent girds for every grid in the first part, thus its label number appears twice in H. There are 3 adjacent grids for every grid in the second part, thus its label number appears three times in H. There are 4 adjacent grids for every grid in the third part, thus its label number appears four times in H. Therefore,

$$H = (abcd)^2(x_1x_2\cdots x_{2m+2n-8})^3(y_1y_2\cdots y_{(m-2)(n-2)})^4$$
$$= (x_1x_2\cdots x_{2m+2n-8})^3.$$

If $x_1x_2\cdots x_{2m+2n-8} = 1$, then $H = 1$, and in this case the number of good couples is even. If $x_1x_2\cdots x_{2m+2n-8} = -1$, then $H = -1$, and in this case the number of good couples is an odd. It shows that whether S is even or odd is determined by the colors of the grids in the second part. Moreover, when the number of blue grids in the second part is odd, then S is odd. Otherwise S is even.

Remark Obviously, we can use the evaluation method in solving example 1 and example 2. Moreover the coloring method can be used in example 5.

Exercise 9

1 Is it possible to cover an 11×12 rectangle with 19 rectangles of 1×6 or 1×7? (Reasoning is required.)

2 When two adjacent unit squares are removed from a row of the (2 row) $\times$ (3 column) rectangle, the obtained shape is called an "L-shape".

(1) Suppose that an $m \times n$ ($m > 1$, $n > 1$) chessboard has a perfect cover with some "L-shapes". Show that mn is divisible by 8.

(2) Find all positive integers m, n ($m > 1$, $n > 1$) such that an $m \times n$ chessboard has a perfect cover with some "L-shapes".

3 Find the smallest positive integer n such that among any n persons, there exist 5 persons who can be divided into two triples with a common member and in every triple, three persons are either all mutually acquainted or all mutually unacquainted.

4 Find the smallest positive integer n such that among any n persons, there exist 4 persons who can be divided into two triples with two common members and in every triple, three persons are either all mutually acquainted or all mutually unacquainted.

5 There are 9 mathematicians at a conference, each speaks at most three languages. For any three mathematicians, at least two persons speak a common language. Show that there are three mathematicians who know a common language. (7th American Mathematical Olympiad)

6 Prove the conclusion of example 1 using the evaluation method.

7 There are two same cogwheels A and B. The cogwheel B is set on a horizontal plane and the cogwheel A is set over the cogwheel B

such that these two cogwheels are coincident. (Thus the projection of two cogwheels is the projection of a whole cogwheel.) Afterwards, we knock off any four pairs of coincident teeth.

If two cogwheels have 14 teeth respectively, is it possible to rotate the cogwheel A to proper position so that the projection of two cogwheels is the projection of a whole cogwheel?

If two cogwheels have 13 teeth respectively, then does the conclusion also hold? Justify your answer. (The Selective Examination for 5th CMO in 1990)

Chapter 10

Reduction to Absurdity and the Extreme Principle

10.1 Reduction to Absurdity

If it is difficult to prove some conclusions directly, then we may consider using the reduction to absurdity. The basic idea of the method is that it leads to a contradiction from the negation of the conclusion and the contradiction implies that the conclusion is true.

Example 1 Let $ABCDE$ be a convex pentagon on the coordinate plane. Each of its vertices are lattice points (i. e. the abscissa and ordinate of the point are all integers.) The five diagonals of $ABCDE$ form a convex pentagon $A_1B_1C_1D_1E_1$ inside of $ABCDE$ (Figure 10.1). Prove that the convex pentagon $A_1B_1C_1D_1E_1$ contains a lattice point on its boundary or within its interior. (26$^\text{th}$ Russia Mathematical Olympiad in 2000)

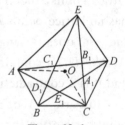

Figure 10.1

Proof Suppose the statement is false. Since the area of any lattice polygon is either an integer or a half-integer, it implying that there is a counterexample to the statement with minimal area. Assume the convex pentagon with minimal area is $ABCDE$. Without loss of generality, assume that

$$S_{\triangle ABC} = \text{Min}\{S_{\triangle ABC}, S_{\triangle BCD}, S_{\triangle CDE}, S_{\triangle DEA}, S_{\triangle EAB}\}.$$

We claim that triangle AC_1D_1 has no lattice point on its edges or its interior, except A. By the assumption, there is no lattice point on $\overline{C_1D_1}$. Suppose, for sake of contradiction, there is a lattice point A'

either on $\overline{AC_1}$ or $\overline{AD_1}$, or the interior of the triangle AC_1D_1. Then pentagon $A'BCDE$ is a convex lattice pentagon with smaller area than the pentagon $ABCDE$. Furthermore, the corresponding inner pentagon is contained within pentagon $A_1B_1C_1D_1E_1$, implying that it does contain a lattice points. But then pentagon $A'BCDE$ has smaller area than pentagon $ABCDE$ and satisfies the same requirements contradicting the minimal definition of pentagon $ABCDE$. By the same reason, triangle CA_1E_1 has no lattice points on its edges or within its interior, except C.

Next, $S_{\triangle ABC} \leqslant S_{\triangle DBC}$ and $S_{\triangle ABC} \leqslant S_{\triangle EAB}$, thus the distance from the point D to the straight line BC is greater than or equal to the distance from the point A to the straight line BC and the distance from the point E to straight line AB is greater than or equal to the distance from the point C to the straight line AB. Therefore the triangle B_1AC contains another vertex O of the parallelogram with two adjacent sides $\overline{BA}$ and $\overline{BC}$ on its edges or within its interior. Since A, B, C are all lattice points, then O is also a lattice point. But the triangle AC_1D_1 has no lattice points on its edges or within its interior, except A and the so does triangle CA_1E_1 except C. Hence the pentagon $A_1B_1C_1D_1E_1$ contains a lattice point O on its boundary or its interior, and it is a contradiction. This completes the proof.

Remark   When we prove the conclusions concerning lattice points, we often start with a convex lattice polygon satisfying required conditions with minimal area.

Example 2   Three vertices of $\triangle ABC$ are colored red, blue and black respectively. The line segments connecting any some interior points and the vertices of $\triangle ABC$ construct many little triangles and the vertices of these little triangles are just these interior points and the vertices of $\triangle ABC$. Suppose every vertex of the little triangles is colored with one of the three colors mentioned. Prove that no matter how to color these vertices, there exists a triangle whose vertices are colored with three different colors.

Proof   Suppose the statement is false. Then there is no triangle whose three vertices are colored with three different colors. The side of the triangle whose two ends are colored with red and blue is called a red-blue side. Assume that the number of all red-blue sides in $\triangle ABC$ is k. Since every red-blue side in $\triangle ABC$ is a common side of two triangles and exactly one of the three sides of $\triangle ABC$ is a red-blue side, the total of red-blue sides in all little triangles is $S = 2k + 1$. On the other hand, suppose that the number of little triangles whose three vertices are red, red and blue or red, blue and blue is P. Since every such triangle has two red-blue sides, and the remaining triangles has no red-blue side (Since there is no triangle whose three vertices are red, blue and black.), $S = 2p$. We obtain $2p = 2k + 1$, i.e. an even = an odd, and it is a contradiction. Hence the statement is true.

Example 3 There are 2004 girls seating at a circular table. They play a deck of 2004 cards. Initially, the girl G has all 2004 cards in her hand. If some girl has at least two cards (more than one card) in her hand, then she must send a card to each of the girls at her left and right. The game is ended if and only if each of 2004 girls only has one card in her hand. Prove that it is impossible to end the game.

Proof Suppose the statement is false. Let 2004 girls be labeled with $1, 2, \ldots, 2004$. If the girl i has a card A after some several times of sending the cards, then we definite that the card A has the floating value i. Assume that the total of the floating values of 2004 cards is S_k after the kth sending cards and the 1st girl has all 2004 cards initially. Thus $S_0 = 2004$. If in the kth sending cards, the 1st girl send a card to each of the 2nd girl and the 2004th girl, the floating value increases

$$(2 - 1) + (2004 - 1) = 2004,$$

i.e.

$$S_k = S_{k-1} + 2004.$$

If in the kth sending cards, the 2004th girl sends a card to each of the 1st girl and the 2003rd girl, the floating value decreases

$$(2004 - 1) + (2004 - 2003) = 2004,$$

i. e.

$$S_k = S_{k-1} - 2004.$$

If in the kth sending cards, the ith girl sends a card to each of the $(i-1)^{th}$ girl and the $(i+1)^{th}$ girl, the floating value increases

$$[i - (i-1)] + [i - (i+1)] = 0 \quad (2 \leqslant i \leqslant 2003),$$

i.e. $S_k = S_{k-1}$. Thus we get S_k is a multiple of 2004 from $S_0 = 2004$. If the game is ended after some several times of sending cards, then the total of the floating values of 2004 cards is

$$S = 1 + 2 + \cdots + 2004 = 1002 \times 2005,$$

S is not a multiple of 2004, and it is a contradiction. Hence it is impossible to end the game.

Remark From the above proof, we know that the conclusion also holds if the number 2004 is changed to any even $2m$ ($m \in \mathbf{N}_+$). The key of the proof is applying the unchanged property that S_k is a multiple of $2m$.

Example 4 Let 2004 points A_1, A_2, ..., A_{2004} be arranged clockwise in a circle. Initially, the labeled number in A_1 is 0, labeled numbers in A_2, A_3, ..., A_{2004} are all 1. The operation is allowed: the player takes any point A_j, if the labeled number in A_j is 1, then labeled numbers a, b and c in points A_{j-1}, A_j and A_{j+1} are changed to $1-a$, $1-b$, $1-c$ simultaneously ($A_0 = A_{2004}$, and $A_{2005} = A_1$). Is it possible that labeled numbers in points A_1, A_2, ..., A_{2004} are all changed to 0?

Solution It is not possible. For sake of contradiction, assume that the labeled numbers in points A_1, A_2, ..., A_{2004} are all changed to 0 after m operations. Since the parity of two numbers $(1-a) + (1-b) + (1-c) = 3 - (a+b+c)$ and $a+b+c$ are different, the parity of the total of labeled 2004 numbers is changed after an operation. But the total of the 2004 labeled numbers is 2003 (odd!) initially, thus m is also odd.

Next, assume that the number of operations starting from the point A_i is x_i ($1 \leqslant i \leqslant 2004$) and the number of changing the labeled number in the point A_i is y_i. Then $\sum\limits_{i=1}^{2004} x_i = m$, $y_j = x_{j-1} + x_j + x_{j+1}$ $(1 \leqslant j \leqslant 2004$, $x_0 = x_{2004}$, $x_{2005} = x_1)$ and $m = y_2 + y_5 + y_8 + \cdots + y_{2003}$. But 668 numbers y_2, y_5, $\ldots$, y_{2003} are all odds, and their sum is even. It contradicts that m is odd. Therefore it is not possible that the labeled numbers in points A_1, A_2, $\ldots$, A_{2004} are all changed to 0.

Remark   From the above proof, we know that the conclusion also holds if the number 2004 is changed to the number $6m$ ($m \in \mathbf{N}_+$).

Example 5   Is it possible that the integer set is divided into three disjoint subsets such that for any integer n, three numbers n, $n - 50$, and $n + 2005$ belong to three distinct subsets respectively?

Solution   It is not possible. For sake of contradiction, assume that the integer set is divided into three disjoint sets such that for any integer n, three numbers n, $n - 50$, and $n + 2005$ belong to three distinct subsets respectively. Assume that sign $m \leftrightarrow k$ expresses m and k belong to the same subset, and $m \xleftrightarrow{\times} k$ expresses m and k belong to distinct subsets and the sign $(p, q, r) \in \mathfrak{S}$ expresses three integers p, q, r belong to three distinct subsets respectively. If we use $n - 50$ or $n + 2005$ to substitute each n in $(n, n - 50, n + 2005) \in \mathfrak{S}$ respectively, then

$$(n - 50, n - 100, n + 1955) \in \mathfrak{S},$$

and

$$(n + 2005, n + 1955, n + 2 \cdot 2005) \in \mathfrak{S}.$$

Thus we obtain $n + 1955 \xleftrightarrow{\times} n - 50$, and $n + 1955 \xleftrightarrow{\times} n + 2005$, and it follows $n + 1955 \leftrightarrow n$. So we know $(n - 50, n - 100, n) \in \mathfrak{S}$ from $(n - 50, n - 100, n + 1955) \in \mathfrak{S}$. If we use $n - 50$ to substitute each n in $(n - 50, n - 100, n) \in \mathfrak{S}$, we know $(n - 100, n - 150, n - 50) \in \mathfrak{S}$. It follows $n \leftrightarrow n - 150$ from $n \xleftrightarrow{\times} n - 50$ and $n \xleftrightarrow{\times} n - 100$. Starting with $n \leftrightarrow 1955$ and $n \leftrightarrow n - 150$, we deduce the following relation.

$0 \leftrightarrow 1955 \leftrightarrow 2 \times 1955 \leftrightarrow \cdots \leftrightarrow 50 \times 1955 = 652 \times 150 - 50 \leftrightarrow 651 \times 150 \leftrightarrow \cdots \leftrightarrow 150 - 50 \leftrightarrow -50$.

But $n \xleftrightarrow{\times} n - 50$, so $0 \xleftrightarrow{\times} -50$, and it is a contradiction.

Therefore no partition of the integer set satisfies the given conditions.

10. 2　The Extreme Principle

Starting with the extreme elements（The greatest number or smallest number; the greatest distance or smallest distance; the greatest area or smallest area; the team（team member）whose score is the greatest or smallest and so on）, we deduce that the conclusion is true or starting with the negation of the conclusion and applying the extreme elements, we deduce a contradiction which implies that the conclusion holds. It is the method of solving problem applying the extreme principle.

Example 6　There are three schools and the number of students of each school is n. Suppose that any student is mutually acquainted with $n+1$ student of the other schools. Prove that there is a student in each school such that the three students are mutually acquainted.

Proof　As there are $3n$ students in the three schools, we choose a student A satisfying that among the students in some of other two schools, the number of students acquainted with A is maximal. Assume this maximal number is k, then $k \leqslant n$. Without loss of generality, assume that this student A belongs to the first school, and he（or she）is mutually acquainted with k students of the second school, then among students in the third school, the number of the persons acquainted with A is $n+1-k \geqslant 1$, i. e. A is acquainted with at least one student C in the third school.

If C is acquainted with some student B in the second school, and B is acquainted with A, then three students A, B, and C are mutually acquainted. Hence the conclusion holds. Otherwise, among the students in the second school, the number of students acquainted with C is at most $n-k$. Therefore among the students in the first school the number of students acquainted with C is at least $n+1-(n-k)=k+1$. It contradicts the above assumption that k is maximal. This completes the proof.

Example 7　In a football tournament, each team played exactly one game against each of the other teams. In each game the winner was awarded 2 points and the loser gets 0 point. If the game is a tie, each of the two teams earned 1 point. Assume that the score of the team A is maximal (the scores of other teams is less than the score of A), but the number of games in which A wins is minimal. Find the smallest value of the number of teams in this tournament. (16th All Russian Mathematical Olympiad)

Solution Suppose that the score of the A team is maximal, and the number of games A wins is n, and the number of games in which A ties its competitor is m, then the total score of A team is $2n + m$. By the given conditions, the number of games other teams wins is at least $n + 1$. Hence the score of each of other teams is at least $2(n + 1)$. and $2n + m > 2(n + 1)$, $m \geq 3$, i. e. A ties at least three teams. Hence there is a team B such that B ties A, and it follows that the score of B is at least $2(n + 1) + 1$, and $2n + m > 2(n + 1) + 1$, $m \geq 4$.

Assume that there are S teams in this tournament, then the A team wins once at least, otherwise, the score of A is at most $S - 1$ and the score of each of other teams is less than $S - 1$. Then the total scores of S teams is less than $S(S - 1)$. But the total scores of S teams is $2\binom{S}{2} = S(S - 1)$, and it is a contradiction. Thus $m \geq 4$, $n \geq 1$, i. e. A team plays at least 5 tournaments, and it follows that the number of teams in this football tournament is at least 6.

	A	B	C	D	E	F	score
A	—	1	1	1	1	2	6
B	1	—	2	0	0	2	5
C	1	0	—	0	2	2	5
D	1	2	2	—	0	0	5
E	1	2	0	2	—	0	5
F	0	0	0	2	2	—	4

On the other hand, as above table, 6 teams A, B, C, D, E, F

satisfy the condition of this problem: the score of the team A is maximal (The scores of other teams is less than the score of A) but the number of games A wins is minimal.

Summarizing what we have described above, the smallest value of the number of teams in this football tournament is 6.

Exercise 10

1 There are n ($\geqslant 5$) members in some committee and $n + 1$ triples satisfying that not all members in any two triples are the same. Prove that there exist two triples with just a common member. (8th American Mathematical Olympiad)

2 On an 8×8 chessboard, there are 9 chessmen occupying a 3×3 square in the bottom left corner of this chessboard and every chessman occupies a 1×1 square. The only one allowed move is jumping over one adjacent occupied square to another unoccupied one (The direction of the jump may be horizontal, vertical or diagonal). Is it possible that the 9 chessmen also occupy a 3×3 square in (1) the top left corner, (2) the top right corner of this chessboard after several jumps?

3 Suppose n (> 1) is odd and $k_1, k_2, \ldots, k_n$ are given positive integers. For every permutation $a = (a_1, a_2, \ldots, a_n)$ of n numbers 1, 2, $\ldots$, n, we denote $S(a) = \sum_{i=1}^{n} k_i a_i$. Prove that there exist two permutations b and c such that $S(b) - S(c)$ is divisible by $n!$. (2nd IMO)

4 There are 17 persons at a conference and each person has just four acquaintances. Prove that there exist two persons such that they are mutually unacquainted and have no common acquaintance.

5 There are n persons at a conference and each person has 8 acquaintances. Suppose that each pair who is mutually acquainted has exactly four common acquaintances and each pair who is mutually unacquainted has exactly two common acquaintances. Find all positive integer n satisfying the given conditions.

6 In a tennis club 20 members play 14 games and each player plays at least one game. Prove that there exist 6 games involving 12 different players (18th American Mathematical Olympiad)

7 There is a football tournament between 20 teams. Find the smallest value of number n of games such that among any three teams there exists a game between two teams after n games.

8 At a volleyball tournament each team played exactly one game against each of the other teams. Assume that the game is not a tie. If the team A wins the team B or the team A wins the team C and the team C wins the team B, then we call that the team A excels the team B. If one team excels all other teams, then this team is called the champion. Is it possible that there exsit exactly two champions in the end?

Chapter 11 Local Adjustment Method

We carry out some adjustments (or some transformations) according to the requirement of the question such that the distance between the existing conclusion and the ultimate objective is less and the aim will be attained step by step. Thus we may rest assured that the final objective is reached through finite adjustments (or finite transformations). This is called the local adjustment method. Using the local adjustment method we can solve the following problems:

(1) To prove that some combinatorial objects have given characteristics;

(2) To solve combinatorial extremum problems;

(3) To prove that there exist some combinatorial objects which have the given properties;

(4) To solve problems about attaining objectives by one's operation.

Example 1 In an $n \times n$ ($n \geqslant 4$) square table each unit square is written with 1 or -1. The product of n numbers in distinct rows and columns is called a basic-term. Let S denote the total of all basic-terms. Prove that for any such table, S is divisible by 4 (China Mathematical Competition in 1989)

Proof Obviously, for any such table, the number of all basic-terms is $n!$. Assume that the unit square in the ith row and the jth column is written with the number a_{ij} ($1 \leqslant i, j \leqslant n, n \geqslant 4$). Since the number of basic-terms containing a_{ij} equals $(n-1)!$, if the sign of the number a_{ij} is changed (i.e. changes from 1 to -1 or from -1 to 1), then the number of the basic-terms whose signs are changed equals $(n-1)!$. Suppose that the number of basic-terms which changes from

-1 to 1 is h, and the number of basic-terms which changes from 1 to -1 is $(n-1)! -h$ and the total of all basic terms is changed from S to S', thus the increment ΔS of S is

$$\Delta S = S' - S = 2h - 2[(n-1)! -h] = 4h - 2(n-1)!.$$

Since $n \geqslant 4$, then $S' - S$ is divisible by 4, i. e. when the sign of some a_{ij} is changed, the increment ΔS of S is the multiple of 4.

If all the numbers a_{ij} in the table are 1, then the total of all basic-terms is $S = n!(n \geqslant 4)$, and it is a multiple of 4. If not all the numbers a_{ij} in the table are 1, then each -1 in this table can be changed to 1 one by one and every time, the increment ΔS of S is a multiple of 4. Hence after several adjustments, all the numbers a_{ij} in the table are changed to 1 and then the total of all basic-terms is $n!$, which is a multiple of 4. Therefore the total of all basic-terms in the original table is a multiple of 4. This completes the proof.

Example 2   Suppose that 2008 is written as the sum of some distinct positive integers such that the product of these positive integers is maximum, find this maximum value.

Solution   Since the number of ways that 2008 can be written as a sum is finite, there exist a case such that the product of these positive integers attains maximum. Suppose that when 2008 is divided into the sum of k distinct positive integers $a_1, a_2, \ldots, a_k$, their product $\prod_{i=1}^{k} a_i$ is maximal. We denote this division by $S = \{a_1, a_2, \ldots, a_k\}$, $(a_1 < a_2 < \cdots < a_k)$. Let $\sum S$ expresses the sum of all numbers in S and $\prod S$ expresses the product of all numbers in S. Thus S has the following properties:

(1) There is at most one positive integer $a \in [a_1, a_2]$ and $a \notin S$. In fact, assume that $a, b \in [a_1, a_k](a < b)$, $a, b \notin S$ and $a-1 \in S$, $b+1 \in S$.

Set

$$S' = (S \setminus \{a-1, b+1\}) \cup \{a, b\},$$

then

$$\sum S' = \sum S = 2008$$

and

$$\frac{\prod S'}{\prod S} = \frac{ab}{(a-1)(b+1)} = \frac{ab}{ab-(b-a)-1} > 1,$$

then $\prod S' > \prod S$.

It contradicts that $\prod S$ is maximal.

(2) $a_1 \neq 1$. In fact, if $a_1 = 1$, let $S' = (S \backslash \{1, a_k\}) \cup \{a_k + 1\}$, then

$$\sum S' = \sum S = 2008$$

and

$$\frac{\prod S'}{\prod S} = \frac{a_k + 1}{1 \cdot a_k} > 1, \quad \prod S' > \prod S.$$

It contradicts that $\prod S$ is maximal.

(3) $a_1 = 2$ or 3.

(i) If $a_1 = 4$ and $5 \in S$, put $S' = (S \backslash \{5\}) \cup \{2, 3\}$, then

$$\sum S' = \sum S = 2008$$

and $\dfrac{\prod S'}{\prod S} = \dfrac{2 \times 3}{5} > 1, \quad \prod S' > \prod S.$

It contradicts that $\prod S$ is maximal.

(ii) If $a_1 = 4$ and $j \notin S$ ($j = 5, 6, \ldots, t-1$), but $t \in S$ ($t \geqslant 6$), so let

$$S' = (S \backslash \{4, t\}) \cup \{2, 3, t-1\},$$

then

$$\sum S' = \sum S = 2008$$

and

$$\frac{\prod S'}{\prod S} = \frac{2 \times 3 \times (t-1)}{4t} = \frac{4t + 2(t-3)}{4t} > 1,$$

then $\prod S' > \prod S$.

It contradicts that $\prod S$ is maximal.

(iii) If $a_1 \geqslant 5$, set $S' = (S \backslash \{a_1\}) \cup \{2, a_1 - 2\}$, then

$$\sum S' = \sum S = 2008$$

and

$$\frac{\prod S'}{\prod S} = \frac{2(a_1 - 2)}{a_1} = \frac{a_1 + (a_1 - 4)}{a_1} > 1,$$

then $\prod S' > \prod S$.

It contradicts that $\prod S$ is maximal.

From (1), (2) and (3), we know that if $a_1 = 3$, then

$$3 + 4 + 5 + \cdots + n - t = 2008,$$

i. e.

$$(n-2)(n+3) = 4016 + 2t.$$

Hence $n = 63$, and $t = 5$, i. e. $S = \{3, 4, 6, 7, \ldots, 63\}$. Set

$$S' = (S \backslash \{7\}) \cup \{2, 5\},$$

then

$$\sum S' = \sum S = 2008$$

and

$$\frac{\prod S'}{\prod S} = \frac{2 \times 5}{7} > 1, \quad \prod S' > \prod S.$$

It contradicts that $\prod S$ is maximal. Therefore there is $a_1 = 2$, i. e. when

$$S = \{2, 3, 4, 5, 6, 8, 9, \ldots, 63\},$$

$\prod S$ attains maximum which equals $\dfrac{63!}{7}$.

Example 3   There are 14 persons playing the "Japanese chess game", and each player plays a game against each of the other 13 players and the game has no tie. For three players A, B and C, if A wins B, B wins C and C wins A, then the triple (A, B, C) is called a "triangle combination". Find the maximum of number of the "triangle combinations". (2002 Japan Mathematical Olympiad)

Solution   Let A_1, A_2, $\ldots$, A_{14} represent 14 persons and A_i win a_i other players ($i = 1, 2, \ldots, 14$). If three players do not form a "triangle combination", then among the three players, there is a player who wins the others. Hence the number of triples which are not the "triangle combinations" is $\sum\limits_{i=1}^{14} \binom{a_i}{2}$ $\left(\text{with the convention } \binom{2}{0} = \binom{2}{1} = 0\right)$ and the number of the "triangle combinations" is $S = \binom{14}{3} - \sum\limits_{i=1}^{14} \binom{a_i}{2}$. Thus S attains maximum if and only if $\sum\limits_{i=1}^{14} \binom{a_i}{2}$ is a minimum. Next, we prove that if $\sum\limits_{i=1}^{14} \binom{a_i}{2}$ attains minimum, then $|a_i - a_j| \leqslant 1 \ (1 \leqslant i < j \leqslant 14)$.

In fact, assume that there exist $a_i - a_j \geqslant 2 \ (i \neq j, 1 \leqslant i, j \leqslant 14)$, and set $a_i' = a_i - 1$, $a_j' = a_j + 1$, $a_k' = a_k \ (k \neq i, j, 1 \leqslant k \leqslant 14)$, then

$$\sum_{i=1}^{14} a_i' = \sum_{i=1}^{14} a_i = \binom{14}{2} = 91,$$

but

$$\sum_{t=1}^{14} \binom{a_t'}{2} - \sum_{t=1}^{14} \binom{a_t}{2}$$

$$= \binom{a_i - 1}{2} + \binom{a_j + 1}{2} - \left(\binom{a_i}{2} + \binom{a_j}{2} \right)$$

$$= - \left(\binom{a_i - 1}{1} - \binom{a_j}{1} \right)$$

$$= - [(a_i - 1) - a_j]$$

$$= - (a_i - a_j) + 1$$

$$\leqslant -2 + 1 \leqslant -1.$$

Hence

$$\sum_{t=1}^{14} \binom{a_t'}{2} < \sum_{t=1}^{14} \binom{a_t}{2},$$

which contradicts that $\displaystyle\sum_{i=1}^{14} \binom{a_i}{2}$ is minimal.

From $\displaystyle\sum_{i=1}^{14} \binom{a_i}{2} = 91$ and $| a_i - a_j | \leqslant 1 (1 \leqslant i, j \leqslant 14)$, we obtain that among $a_1, a_2, \ldots, a_{14}$, there are seven 6s and seven 7s, thus the minimal value of $\displaystyle\sum_{i=1}^{14} \binom{a_i}{2}$ equals

$$7 \binom{6}{2} + 7 \binom{7}{2} = 252$$

and the maximal value of

$$S = \binom{14}{3} - \sum_{i=1}^{14} \binom{a_i}{2}$$

equals $\binom{14}{3} - 252 = 112$.

On the other hand, when $1 \leqslant i \leqslant 7$, let A_i win $A_{i+1}, A_{i+2}, \ldots,$ A_{i+6} and $A_{i+7}, A_{i+8}, \ldots, A_{i+13}$ win A_i (with convention $A_{j+14} = A_j$), and when $8 \leqslant i \leqslant 14$, let A_i win $A_{i+1}, A_{i+2}, \ldots, A_{i+7}$ and $A_{i+8},$ $A_{i+8}, \ldots, A_{i+13}$ win A_i (with the convention $A_{j+14} = A_j$), then in this case, the equality $S = c - \displaystyle\sum_{i=1}^{14} \binom{a_i}{2} = 112$ holds.

Summarizing what we have described above, we obtain that the maximal value of the number of the " triangle combinations" equals 112.

Example 4   Suppose that several heaps of little balls are laid around a circle and the numbers of balls in each heap are all the multiples of 3 (the numbers of balls in each heap may be different). Each time, the following adjustment is allowed: Each heap is divided into three equal parts and one part is remained while the other two parts are sent to the left and right heaps respectively. If the number of balls in some heap is not a multiple of 3, then we take one or two little balls from the bag which is supplied to this heap such that the number of balls in this heap is a multiple of 3. Then, the adjustment is similar. Is it possible that with several adjustments, the number of balls in each heap is equal?

Solution   Suppose that the maximum of the numbers of balls in the heaps is $3m$ and the minimum is $3n$ before some adjustment, so $m > n$. then the following conclusions hold.

(1) After an adjustment, the number of balls in each heap is also less than or equal to $3m$ and more than or equal to $3n$.

In fact, suppose that there are $3l$ balls in some heap, and there are $3k$ and $3h$ balls in its left and right heaps respectively ($3m \geqslant 3l \geqslant 3n$, $3m \geqslant 3k \geqslant 3n$, $3m \geqslant 3h \geqslant 3n$). Thus after an adjustment, the number of little balls in this heap equals $l + k + h$ satisfying that $3m \geqslant l + k + h \geqslant 3n$. If $l + k + h$ is not a multiple of 3 (i.e. $3m > l + k + h > 3n$), then we supply one or two balls to this heap such that the number of balls in this heap is a multiple of 3 which satisfies the number of balls in this heap is less than or equal to $3m$ and more than or equal to $3n$.

(2) If the number of balls in some heap is more than $3n$, then after an adjustment, the number of balls in this heap is also more than $3n$.

In fact, as (1), if $3l > 3n$, $3k \geqslant 3n$, and $3h \geqslant 3n$, then $l + k + h > 3n$.

(3) There is at least one heap in which the number of balls equals $3n$, after an adjustment, the number of balls in this heap is more than $3n$.

In fact, there is a heap in which the number of balls equals $3n$, the number of balls in the heap on its left or right is more than $3n$. The signs is the same as (1), without loss of generality, and assume that $3k > 3n$, $3l = 3n$, and $3h \geqslant 3n$, then $l + k + h > 3n$.

Hence after an adjustment, the number of heaps in which the number of balls equals $3n$ is decreasing. Therefore with several adjustments, the numbers of balls in each heap will be more than $3n$. But all the numbers of balls in the heaps are less than or equal to $3m$, thus the difference f between the maximum and minimum of the numbers of balls in these heaps is decreasing. Consequently, with several adjustments, this result must be $f = 0$, i. e. the numbers of balls in these heaps are equal.

Remark   To solve problems about attaining objects by one's operation, we often construct a objective function f whose value is a nonnegative integer. If the object is not attained, then we may carry out some proper adjustments such that the value of the objective function f is decreasing. Since the value of the objective function f is a nonnegative integer, the process of adjustments can not be infinite. We have proved that with several adjustments, the object must be attained. In this problem, the objective function is $f = 3m - 3n$.

Example 5   There are at least four candy bars in n ($n \geqslant 4$) boxes. Each time, one is allowed to take a candy from each of any two boxes and put the two candies to the third box. Is it possible that all the candies are in a box after several operations. (9th CMO in 1994)

Solution I   It is always possible to all the candy bars into one box. We will prove our statement by induction on m, the number of candy bars. For the base case $m = 4$, there are at most 4 nonempty boxes. We disregard all other empty boxes and consider all possible initial distributions:

(1) (1, 1, 1, 1), (2) (1, 2, 1, 0), (3) (2, 2, 0, 0), (4) (1, 3,

$0, 0)$.

For distribution (1)-(4), we proceed as follows:

(1) $(1, 1, 1, 1) \to (3, 1, 0, 0) \to (2, 0, 2, 0) \to (1, 2, 1, 0) \to$ $(0, 4, 0, 0)$;

(2) $(1, 2, 1, 0) \to (0, 4, 0, 0)$;

(3) $(2, 2, 0, 0) \to (1, 1, 2, 0) \to (0, 0, 4, 0)$;

(4) $(1, 3, 0, 0) \to (0, 2, 2, 0) \to (2, 1, 1, 0) \to (4, 0, 0, 0)$.

Thus the base case is proven.

Now we assume that the statement is true for some positive integer $m \geqslant 4$. If we give $m + 1$ candy bars, we mark one of them and called it special. We first ignore the special candy bar and just consider the other m candies. By the induction hypothesis, we can put all m candy bars into one box. If this box also contains the special candy bar, we are done. If not, we put two empty boxes and proceed as follows:

$(1, m, 0, 0) \to (0, m-1, 2, 0) \to (2, m-2, 1, 0) \to (2, m-3, 0, 2) \to (1, m-1, 0, 1) \to (0, m+1, 0, 0)$.

Now all candy bars are in one box and our induction is complete.

Solution II   We first prove that through finite operation, we can put all candy bars into two or three boxes.

In fact, if the number of nonempty boxes is at least three. We take any three nonempty boxes A, B, C and assume there are a, b, c $(0 < a \leqslant b \leqslant c)$ candy bars in the three boxes respectively and proceed as follows:

$$(a, b, c) \to (a-1, b-1, c+2) \to (a-2, b-2, c+4)$$
$$\overset{a-2 \text{ times}}{\to} \cdots \to (0, b-a, c+2a),$$

i.e. the numbers of nonempty boxes decrease one at least. Thus with several operations we can put all candy bars into two or three boxes.

Next, without loss of generality, we assume that all candy bars are put into three boxes A, B, and C and there are a, b, c $(0 \leqslant a \leqslant b \leqslant c)$ candy bars in A, B and C respectively. We take an empty box (Since $n \geqslant 4$, this empty box is existential) and denote this distributions by $(a, b, c, 0)$. If two numbers in a, b and c are equal, then from the above proof, we know that with several operations, we can put all candy bars into one box. Hence we just consider the case: $a > b > c \geqslant 0$. It is also divided into the two following cases:

(1) If $a = c + 2$, then $b = c + 1$, and $a + b + c = 3c + 3 \geqslant 4$, i. e. $c \geqslant 1$. We proceed as follows:

$$(c + 2, c + 1, c, 0) \rightarrow (c + 1, c, c, 2) \rightarrow (c, c + 2, c, 1) \rightarrow$$

$$(c - 1, c + 2, c + 2, 0) \rightarrow (c + 1, c + 1, c + 1, 0) \xrightarrow{c+1 \text{ times}} \cdots \rightarrow$$

$$(3c + 3, 0, 0, 0).$$

(2) If $a > c + 2$, we carry out an operation as follows: $(a, b, c) \rightarrow (a - 1, b - 1, c + 2)$. Since $a > b > c$ and $a > c + 2$, then $a - 1 > b - 1 \geqslant c$, and $a - 1 \geqslant (c + 3) - 1 = c + 2$. The maximum of numbers of candy bars in three boxes is decreasing but the minimum is not decreasing. Therefore with several operations, we can get one of the two following cases: the numbers of candy bars in two boxes are equal or as case (1). From the above proof, we know that with several operations, we can put all candy bars into one box.

Exercise 11

1 There are $2n$ points on the line segment AB and they are symmetrical about the midpoint M of AB. Among these points, n points are colored red and the other n points are colored blue. Prove that the sum of distances from the point A to all red points equals the sum of distances from the point B to all blue points

2 We add the sign " $+$ " or " $-$ " to each of the numbers $1, 2, \ldots,$ 2005 such that their algebraic sum is nonnegative and reaches the minimum. Find the expression.

3 There are 1989 points in space, no three of which are collinear. They are divided into 30 groups such that the numbers of elements in any two groups are distinct. We take a point from each of any three groups which determine a triangle. Find the numbers of elements in each group such that the number of these triangles reaches maximum. (The 4th CMO)

4 If an $m \times m$ square could be divided into 7 rectangles satisfying that any two of them have no common interior point and the lengths of 14 sides in these rectangles are 1, 2, 3, 4, 5, 6, 7, 8, 9, 10, 11, 12, 13 and 14. Find the maximum of positive integer m.

5 Consider the permutation $(a_1, a_2, \ldots, a_{20})$ of numbers 1, 2, ..., 20. For this permutation, we can swap any two numbers. The ultimate aim is that with several swaps $(a_1, a_2, \ldots, a_{20})$ changes to $(1, 2, \ldots, 20)$. Suppose that for each permutation $a = (a_1, a_2, \ldots, a_{20})$, we denote the minimum of swaps to achieving the aim with k_a. Find the maximum of k_a.

6 There are $2n + 1$ players $p_1, p_2, \ldots, p_{2n+1}$ in a table tennis tournament, and each player plays a game against each of the other $2n$ players and the game has no tie. Suppose that the player p_i wins w_i times. Find the maximum and minimum of $S = \sum_{i=1}^{2n+1} w_i^2$.

Chapter 12   Construction Method

To prove that some statements in combinatorial mathematics are true, we often use the construction method. The construction methods are divided into two classes: the direct construction method and inductive construction method. If it is difficult to construct a object satisfying the given conditions, then we may start with the following aspects:

(1) We analyze the structures of the object which must be constructed by us and then we will construct this object satisfying all conditions according to these structure (the combinatorial analysis method);

(2) We construct some parts satisfying partial conditions and then the object which must be constructed by us will consist of these parts (the method of consisting of parts);

(3) We construct an object satisfying parts of required conditions and then we carry out some adjustments step by step such that the object satisfies all conditions (the method of stepwise adjustment).

If some object concerning a positive integer n must be constructed by us, and it is difficult to construct this object directly, we could construct it by using the inductive construction method.

Example 1   Is there a set M_0 of infinitely many circles in the rectangular coordinates system satisfying the following conditions?

(1) Any two circles in M_0 and their interiors have at most one common point;

(2) Each rational point in the x-axis is on some circle in M_0.

Solution   For any rational points $(r, 0)$ in the x-axis, where $r =$

$\frac{p}{q}$ (p and q are two coprime integers, $q > 0$), we draw a circle above the x-axis with the radius $R_r > 0$ such that this circle and the x-axis are tangent at the point $(r, 0)$ and we denote this circle by $C_r(R_r)$. The set consisting of these circles is denoted by M_0. Obviously, the set M_0 satisfies the condition (2). To make the set M_0 satisfy the condition (1), we analyze the conditions which must be satisfied by R_r.

Let $C_{r_i}(R_{r_i})$ ($i = 1, 2$) be any two circles in the set M_0, because the two circles are above the x-axis and they are tangent to the x-axis, the two circles intersect if and only if their central distance is less than the sum of the radii of two circles, i.e.

$$\sqrt{(r_1 - r_2)^2 + (R_{r_1} - R_{r_2})} < R_{r_1} + R_{r_2} \Leftrightarrow (r_1 - r_2)^2 < 4R_{r_1} R_{r_2}.$$

Setting $r_i = \frac{p_i}{q_i}$ (p_i and q_i are two coprime integers and $q_i > 0$, $i = 1, 2$), we obtain

$$(p_1 q_2 - p_2 q_1)^2 < 4q_1^2 q_2^2 R_{r_1} R_{r_2}. \qquad \text{①}$$

If we take $R_{r_i} = \frac{1}{kq_i^2}$ ($k \geqslant 2$), then from ① we obtain

$$(p_1 q_2 - p_2 q_1)^2 < \frac{4}{k^2} \leqslant 1.$$

If $r_1 \neq r_2$, then the left part of above inequality is a positive integer, and it is a contradiction. Therefore as long as we take $R_r = \frac{1}{kq^2}$ ($k \geqslant 2$), the set M_0 satisfies all the conditions of the problem. Thus we have proved that there exist a set with infinitely many circles:

$$M_0 = \left\{ C_r(R_r) \,\middle|\, r = \frac{p}{q}, \ R_r = \frac{1}{kq^2} \ (k \geqslant 2), \ p, \right.$$

$$\left. \text{and } q \text{ are two coprime integers, } q > 0 \right\}$$

satisfying the conditions (1) and (2) of the problem. Obviously the number of such set M_0 is infinite.

Example 2 Is it possible that the set N_+ of positive integers is divided into two disjoint sets A and B satisfying the following conditions:

(1) Any three distinct numbers of A cannot form an arithmetic sequence.

(2) Any infinite numbers of B cannot form an infinite arithmetic sequence with the common difference $d \neq 0$?

Analysis Let $A = \{a_1, a_2, a_3, \ldots\}$ $(a_1 < a_2 < a_3 < \cdots)$ and $B = N_+ \setminus A$ satisfy the above conditions (1) and (2). If three distinct numbers a, b, $c'(a < b < c)$ of A form an arithmetic sequence, then $2b = a + c > c$. Thus as long as $a_{i+1} \geqslant 2a_i$ $(i = 1, 2, 3, \ldots)$, any three numbers of A cannot form an arithmetic sequence, and it is not difficult to construct a such set A. To make the set B satisfies the above condition (2), we just need to let at least one term of the infinite arithmetic sequence $\{a + nd\}$ $(n = 0, 1, 2, \ldots)$ belong to the set A, for any a, $d \in N_+$.

Solution Let (a, d) denote the infinite arithmetic sequence with the first term a and the common difference d. We arrange all arithmetic sequences of positive integers with the common difference $d \neq 0$ in following order: $(1, 1)$, $(1, 2)$, $(2, 1)$, $(1, 3)$, $(2, 2)$, $(3, 1)$, $\ldots$, according to the following rule: If $a + d < a' + d'$ or $a + d = a' + d'$ and $a < a'$, then (a, d) is arranged in front of (a', d').

Thus we construct the sequence as follows: Let $a_1 = 1$, if a_1, $a_2, \ldots, a_n$ have been chosen, then among $(n + 1)^{\text{th}}$ arithmetic sequence, we take a number which is greater than $2a_n$ as a_{n+1}. Let $A = \{a_1, a_2, \ldots, a_n, \ldots\}$, since $a_{n+1} > 2a_n$, then any three numbers of A cannot form an arithmetic sequence. Next, we set $B = N_+ \setminus A$, and there are at least one term in each of infinite positive integer arithmetic sequences with the common difference $d \neq 0$ belonging to A. Thus any infinite numbers of B cannot form an infinite arithmetic sequence with the common difference $d \neq 0$. This prove that there exist two sets A and B satisfy the conditions (1) and (2) of the problem.

Example 3   Is there a finite set M of points in the plane satisfying for each point P of M, there exist just three points of M such that the distances from these three points to P are proximate?

Analysis   Let M_0 be a set of points in the place. Assume for the point $P \in M_0$, there exist exactly three points of M_0 such that the distances from these three points to P are proximate, then the point P is called a good point of set M_0. As in figure 12.1, the rhombus consisting of two regular triangles has four vertices, the set M_0 consisting of the four vertices just has two good points, thus this set M_0 does not satisfy the condition of the problem.

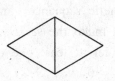   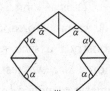

Figure 12. 1   **Figure 12. 2**

Next, as in figure 12.2, we will construct a figure G which consists of m such rhombuses and m line segments such that it satisfies the condition of the problem. (The length of side of the rhombuses equals the length of each of these line segments.) Each point in this figure G is a good point as long as

$$90° < \alpha \leqslant 120°. \tag{①}$$

By the formula of the sum of interior angles of a convex polygon, we obtain that

$$m \cdot 120° + 2m(\alpha + 60°) = (3m - 2) \cdot 180° \Rightarrow \alpha = 150° - \frac{180°}{m}. \tag{②}$$

Substituting ② to ①, we know $3 < m \leqslant 6$. Thus $m = 4$, 5 or 6. So $\alpha = 105°$, if $m = 4$; $\alpha = 114°$, if $m = 5$ or $\alpha = 120°$, if $m = 6$.

Solution   Three sets of points (figure 12.3) all satisfy the condition of the problem. (In figure 12.3, for any point P, we use solid lines to connect P and the three points proximate P.)

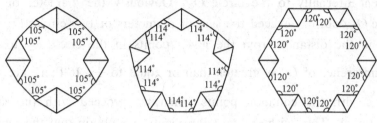

Figure 12. 3

Remark In this solution, the method of consisting of parts is used. Obviously the unions of some such sets as in figure 12. 3 all satisfy the conditions of the problem as long as the minimal distance between any two points from such sets is greater than the minimal distance between any two points of each set as in figure 12. 3. Moreover when we use the construction method to solve an existence problem, we often give the solution and omit the process of designing and exploring of the construction.

Example 4 Assume that there are 2000 points in the plane. Prove that the 2000 points could be covered by some circular figures satisfying the following conditions:

(1) The sum of the diameters of these circles does not exceed 2000;

(2) The distance between any two circle figures is greater than 1.

Proof Firstly, we prove that there are several circular figures satisfying the condition (1). In fact, we take 2000 circular figures with the diameter 1 such that the center of each circle is a given point. Thus the 2000 circular figures cover 2000 given points and the sum of diameters of these circles is 2000.

Next, if two circular figures have the common points, as in figure 12. 4, then we carry out an adjustment. As in figure 12. 4, we use a bigger circular figure O_3 take the place of the two circular figures O_1 and O_2, and the centers of circles O_3, O_1 and O_2 are collinear and two circles O_1 and O_2 are

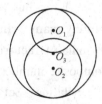

Figure 12. 4

tangent internally to the circle O_3. Obviously the diameter of the circle O_3 does not exceed the sum of diameters of two circles O_1 and O_2 and the distance from the given points in the circle O_3 to the circumference of O_3 is greater than or equal to $\frac{1}{2}$. If there are two circles with the common points, then we proceed with the same adjustment. Thus with several adjustments, we obtain that the sum of diameters of the circular figures without common point does not exceed 2000 and they cover the given 2000 points and the distance from each point to the circumference of the circular figure which covers this point is greater than or equal to $\frac{1}{2}$. Let the minimum of distances between any two circular figures be d, then $d > 0$.

If $d > 1$, then the conclusion holds. If $0 < d \leqslant 1$, then we proceed with an adjustment as follows: we use the circular figure with the same center whose radii is reduced by $\frac{1}{2} - \frac{d}{3}$ to take the place of the original circular figure. Thus these new circular figures cover the 2000 given points and the sum of the diameters of them does not exceed 2000 and the distance between any two of them is greater than or equal to

$$d + 2\left(\frac{1}{2} - \frac{d}{3}\right) = \frac{1}{3}d + 1 > 1.$$

This completes the proof.

Example 5 Is there an infinite sequence of positive integers $a_1 < a_2 < a_3 < \cdots$ such that for any integer A, there are at most finitely many primes in the sequence $\{a_n + A\}_{n=1}^{\infty}$.

Analysis If $|A| \geqslant 2$, then $n! + A$ is a composite number when $n \geqslant |A|$. Thus when $|A| \geqslant 2$, we set $a_n = n!$, then there are at most finitely many primes in the sequence $\{a_n + A\}_{n=1}^{\infty}$. If $A = \pm 1$, by the formula of factorization, we take $a_n = (n!)^3$, then $a_n + A$ is also a composite number,

Solution Set $a_n = (n!)^3$. If $A = 0$, then there is not a prime in the sequence of numbers $\{a_n\} = \{(n!)^3\}$. If $|A| \geqslant 2$, then $a_n + A =$

$(n!)^3 + A$ $(n = 1, 2, 3, \ldots)$ is a composite number (the multiple of A) when $n \geqslant |A|$. If $A = \pm 1$, then

$$a_n + A = (n!)^3 \pm 1 = (n! \pm 1) \times [(n!)^2 \mp n! + 1]$$

is a composite number when $n \geqslant 3$. Hence for any integer A, there are at most finitely many primes in the sequence $\{a_n + A\}_{n=1}^{\infty}$.

Example 6 Prove that for any positive integer $n \geqslant 2$, there exist n distinct positive integers a_1, a_2, a_3, $\ldots$, a_n such that for any $1 \leqslant i < j \leqslant n$, $a_i + a_j$ is divisible by $a_i - a_j$.

Analysis We know the conclusion is true when $n = 2$. Assume that when $n = k$, there exist k positive integers a_1, a_2, a_3, $\ldots$, a_k satisfying the condition of the problem. When $n = k + 1$, set $b_i = a_i + \lambda$ $(1 \leqslant i \leqslant k)$, and $b_{k+1} = \lambda$ (λ is undetermined). To make $b_i + b_j = a_i + a_j + 2\lambda$ be divisible by $b_i - b_j = a_i - a_j$, it is necessary that λ is divisible by $a_i - a_j (1 \leqslant i < j \leqslant k)$ and to make $b_i + b_{k+1} = a_i + 2\lambda$ be divisible by $b_i - b_{k+1} = a_i (1 \leqslant i \leqslant k)$, it is necessary that λ is divisible by a_i. Hence the proof is complete, if we set

$$\lambda = \prod_{i=1}^{k} a_i \prod_{1 \leqslant i < j \leqslant k} (a_i - a_j).$$

Proof When $n = 2$, we easily know the conclusion is true. Assume that when $n = k$, there exist k positive integers a_1, a_2, a_3, $\ldots$, a_k such that for any $1 \leqslant i < j \leqslant k$, $a_i + a_j$ is divisible by $a_i - a_j$. When $n = k + 1$, set $b_i = a_i + \lambda$ $(1 \leqslant i \leqslant k)$, and $b_{k+1} = \lambda$ where

$$\lambda = \prod_{i=1}^{k} a_i \prod_{1 \leqslant i < j \leqslant k} (a_i - a_j),$$

then if $1 \leqslant i < j \leqslant k$, $b_i + b_j = a_i + a_j + 2\lambda$ is divisible by $b_i - b_j = a_i - a_j$. When $1 \leqslant i \leqslant k$, $b_i + b_k = a_i + 2\lambda$ is divisible by $b_i - b_k = a_i$. The proof is complete. Hence for $n \geqslant 2$, there exist n distinct positive integers a_1, a_2, a_3, $\ldots$, a_n such that for any $1 \leqslant i < j \leqslant n$, $a_i + a_j$ is divisible by $a_i - a_j$.

Example 7 Is there a set M satisfying the following conditions:

(1) There are exactly 2008 positive integers in the set M;

(2) Each number of M and the sum of any numbers of M can be represented with the form m^k (m, $k \in \mathbf{N}_+$ and $k \geqslant 2$).

Analysis Obviously the set $\{1, 2, 3, \ldots, 2008\}$ does not satisfy the condition (2). We find a positive integer d such that $M = \{d, 2d, 3d, \ldots, 2008d\}$ satisfies the condition (2). Since each number of M and the sum of any numbers of M belong to the set

$$S = \left\{ d, 2d, \ldots, \frac{2008 \times 2009}{2} d \right\},$$

we just need to find a positive integer d such that each number of S is represented with the form m^k (m, $k \in \mathbf{N}_+$ and $k \geqslant 2$). Generally, we use n instead of $\dfrac{2008 \times 2009}{2}$ and construct the set S satisfying the above conditions by the inductive construction method.

Solution Firstly, we prove the following lemma.

Lemma For any positive integer n, there is a positive integer d_n such that each number of $M = \{d_n, 2d_n, \ldots, nd_n\}$ could be represented with the form m^k (m, $k \in \mathbf{N}_+$ and $k \geqslant 2$).

Proof of Lemma When $n = 1$, we know that the conclusion holds when $d_1 = 3^2$. Assume that for $n \in \mathbf{N}_+$, the conclusion holds, i. e. there exists a positive integer d_n such that $id_n = m_i^{k_i}$ (m_i, $k_i \in \mathbf{N}_+$ and $k_i \geqslant 2$, $i = 1, 2, \ldots, n$). Set $d_{n+1} = d_n [(n+1)d_n]^k$, where k is the common multiple of $k_1, k_2, \ldots, k_n$ and let $k = k_i q_i$ ($1 \leqslant i \leqslant n$), then

$$\begin{aligned}
id_{n+1} &= id_n [(n+1)d_n]^k \\
&= m_i^{k_i} [(n+1)d_n]^{k_i q_i} \\
&= \{ m_i [(n+1)d_n]^{q_i} \}^{k_i} \quad (1 \leqslant i \leqslant n)
\end{aligned}$$

and

$$(n+1)d_{n+1} = [(n+1)d_n]^{k_i+1}.$$

Hence the lemma is proved.

Now we come back to the original problem. Let $n_0 = \dfrac{2008 \times 2009}{2}$, with the lemma we obtain that there exists a positive integer d_{n_0} such that the each number of set $S_{n_0} = \{d_{n_0}, 2d_{n_0}, \ldots, n_0 d_{n_0}\}$ *can be represented with the form* m^k (m, $k \in \mathbf{N}_+$, and $k \geqslant 2$). Thus the set $M = \{d_{n_0}, 2d_{n_0}, \ldots, 2008d_{n_0}\}$ satisfies the conditions (1) and (2) of the problem.

Remark In this problem, the positive integer d_{n+1} is found according to the following steps: when $i = 1, 2, \ldots, n$, to apply the induction hypothesis, d_{n+1} must be a multiple of d_n and $(n+1)d_{n+1}$ can be represented with the form m^k when d_{n+1} has the form $d_n[(n+1)d_n]^k$ and $id_{n+1} = m_i^{k_i}[(n+1)d_n]^k$ ($i = 1, 2, \ldots, n$) can be represented with the form m^{k_i} when k is a multiple of k_i ($i = 1, 2, \ldots, n$), i.e. k is the common multiple of $k_1, k_2, \ldots, k_n$.

Exercise 12

1 Prove that we can fill some distinct perfect square numbers in each grid of the $m \times n$ rectangle such that there is exactly a perfect square number in each grid and the sum of all numbers in each row and column are all perfect squares

2 For any positive integer n, are there n points, not all of which are collinear in the plane such that the distances between any two points are positive integers?

3 When a unit square is moved from a 2×2 square, the obtained shape is called an"L-shape". Is it possible that the following $m \times n$ rectangle is divided perfectly into many"L-shapes"?

(1) $m \times n = 2003 \times 2005$; (2) $m \times n = 2005 \times 2007$.

4 There are four straight lines in a plane, and any two lines are intersecting and no three lines are intersecting at a point. Thus three intersection points determine two line segments on each straight line. Is it possible that the lengths of these eight line segments are (1) 1, 2, 3, 4, 5, 6, 7, 8? or (2) distinct positive integers?

5 Prove that the set of all positive integers can be partitioned into 100 disjoint nonempty subsets such that if three positive integers

a, b, c satisfy $a + 99b = c$, then at least two of them belong to the same subset. (26th Russia Mathematical Olympiad in 2000)

6 Prove that there exist infinitely many positive integers n, such that the set $S_n = \{1, 2, \ldots, 3n\}$ can be partitioned into three disjoint subsets $A = \{a_1, a_2, \ldots, a_n\}$, $B = \{b_1, b_2, \ldots, b_n\}$ and $C = \{c_1, c_2, \ldots, c_n\}$ such that for any $i = 1, 2, \ldots, n$, $a_i + b_i = c_i$ holds.

7 For any positive integer n, is there a finite set M of points in the plane such that for each point $P \in M$, there are exactly n points $A_1, A_2, \ldots, A_n \in M$ such that the distance between P and A_i ($i = 1, 2, \ldots, n$) is equal to 1.

8 Prove that every positive integer n, except a finite number of them, can be represented as a sum of 2004 positive integers: $n = a_1 + a_2 + \cdots + a_{2004}$, where $a_1 < a_2 < \cdots < a_{2004}$ and $a_i \mid a_{i+1}$ ($i = 1, 2, \ldots, 2003$). (2004 CMO)

Chapter 13 — Combinatorial Counting Problems

The combinatorial counting problems are common in the mathematical competition, and the basic methods of solving combinatorial counting problems are as follows:

(1) Enumeration method (The part behind Example 1 in Chapter 5);

(2) Applying the fundamental counting principles and formulas (Examples 1 – 7 in Chapter 1 and Examples 1, and 3 in Chapter 5);

(3) Pairing method, mapping method and general corresponding method (Examples 1 – 2, 9 – 11 in Chapter 6);

(4) Counting in two ways (Examples 1 – 7 in Chapter 7);

(5) Recurrence method (Examples 1 – 5 in Chapter 8);

(6) Applying the inclusion – exclusion principle (Examples 9 – 12 in Chapter 1);

(7) Applying the formula of number of nonnegative integer solutions of the indefinite equation $x_1 + x_2 + \cdots + x_k = n$ (Examples 7 (Solution Ⅱ) and 8 in Chapter 1);

(8) Applying the generating function (Examples 4 – 5 in Chapter 3).

For more complex combinatorial counting problems, sometimes, we may apply several distinct methods simultaneously to find their solutions, and for some other problems, we may give several distinct solutions with different methods.

Example 1   Each pair of the vertices of a cube determines a straight line. Among these straight lines, how many pairs of noncoplanar lines which are not perpendicular are there? (Mathematical summer Camp of Hunan Province of China in 2000)

Solution   If two noncoplanar straight lines are not perpendicular, then they are called a good pair of straight lines. As 1 edge and 6 straight lines (2 main diagonal lines and 4 facet diagonal lines) form the good pairs of straight lines, the number of good pairs of 12 edges is equal to $12 \times 6 = 72$. Next, 1 facet diagonal line and 8 straight lines (4 edges and 4 facet diagonal lines) form the good pairs of straight lines, thus the number of good pairs of 12 facet diagonal lines is equal to $12 \times 8 = 96$. Finally, 1 main diagonal line and 6 straight lines (6 edges) form the good pairs of straight lines, thus the number of good pairs of 4 main diagonal lines is equal to $6 \times 4 = 24$.

Since each pair is counted twice in the above counting, then the number of required good pairs equals

$$\frac{1}{2}(72 + 96 + 24) = 96.$$

Example 2 Let a_1, a_2, $\ldots$, a_{100} be any permutation of $\{1901, 1902, \ldots, 2000\}$. Given such a permutation, we form the sequence of partial sums:

$$S_1 = a_1, \ S_2 = a_1 + a_2, \ S_3 = a_1 + a_2 + a_3, \ \cdots,$$
$$S_{100} = a_1 + a_2 + \cdots + a_{100}.$$

How many of these permutations will contain no terms of the sequence S_1, S_2, $\ldots$, S_{100} that is divisible by three? (32nd Canadian Mathematical Olympiad in 2000)

Solution Let

$$M = \{1901, 1902, \cdots, 2000\} = K_0 \cup K_1 \cup K_3,$$

where $K_i = \{n \mid n \in M \text{ and } n \equiv i \pmod 3\}$ $(i = 0, 1, 2)$, then $|K_0| = |K_1| = 33$, $|K_2| = 34$. Let $a_i' \equiv a_i \pmod 3$, $0 \leqslant a_i' \leqslant 2$ $(i = 1, 2, \ldots, 100)$, thus the permutations $S' = \{a_1', a_2', \ldots, a_{100}'\}$ determine whether the permutations $S = \{a_1, a_2, \ldots, a_n\}$ will have the given property. Since S' contains 33 of the number 0, 33 of the number 1 and 34 of the number 2, if each of partial sums S_1, S_2, $\ldots$, S_{100} is not divisible by 3, then the sequence consisting of 67 elements 1 or 2 of S' must be 1, 1, 2, 1, 2, 1, 2, $\ldots$, 1, 2 or 2, 2, 1, 2, 1, 2, 1, $\ldots$, 2, 1. But $|K_2| = |K_1| + 1$, hence only the second case is possible. Since each 0 could be put in any position except $a'_1 (\neq 0)$, the number of ways to arrange 33 of the number 0 equals $\binom{99}{33}$. Therefore, the number of permutation satisfying conditions equals

$$\binom{99}{33} \cdot 33! \cdot 33! \cdot 34! = \frac{99! \cdot 33! \cdot 34!}{66!}.$$

Example 3 We use six distinct colors to color the six side facets of a cube such that each facet is colored with one of 6 colors and two facets with a common edge must be colored with different colors. How many distinct coloring ways are there? (**Remark** A coloring is the same as another which is from the rotation or rolling of the former.) (China Mathematical Competition in 1996)

Solution Since three facets with a common vertex must be colored with three distinct colors, at least three colors are used. It is divided into the four following cases:

(1) When 6 colors are used, we upturn a facet colored with some color. We choose a color from the remaining colors to color the base facet in $\binom{5}{1}$ ways. Afterward we use the remaining colors to color four other facets in $(4-1)!$ ways. Therefore when 6 colors are used, the number of distinct coloring ways equals $\binom{5}{1} \cdot (4-1)! = 30.$

(2) When 5 colors are used, the number of ways to take 5 colors from 6 colors is $\binom{6}{5}$. In this case, there is a pair of opposite facets are colored with the same color. We choose a color from 5 colors to color a pair of opposite facets in $\binom{5}{1}$ ways. Afterward we use the remaining colors to color four other facets in $\frac{1}{2} \cdot (4-1)!$ ways (it is the number of the necklaces containing 4 distinct beads). Therefore when 5 colors are used, the number of distinct coloring ways equals

$$\binom{6}{5} \cdot \binom{5}{1} \cdot \frac{1}{2} \cdot (4-1)! = 90.$$

(3) When 4 colors are used, the number of ways to take 4 colors from 6 colors is $\binom{6}{4}$. In this case, each of two pairs of opposite facets is colored with the same color and the third pair of facets are colored with distinct colors. We choose two colors from 4 colors to color the third pair in $\binom{4}{2}$ ways. Afterwards we use the remaining colors to color four other facets such that each of two pairs of opposite facets is colored with same color, and there is only one way (it is the number of the necklace containing 2 distinct beads). Therefore when 4 colors are used, the number of distinct coloring ways equals $\binom{6}{4} \cdot \binom{4}{2} \cdot 1 = 90$.

(4) When 3 colors are used, the number of ways to take 3 colors from 6 colors is $\binom{6}{3}$. In this case, each of three pairs of opposite facets is colored with the same color. Afterward we use the 3 colors to color these three groups the three pairs of opposite facets such that each pair is colored with the same color, and there is only one way. Therefore when 3 colors are used, the number of distinct coloring ways equals $\binom{6}{3} \cdot 1 = 20$.

Summarizing what we have described above, we get that the number of distinct colored ways equals $30 + 90 + 90 + 20 = 230$.

Example 4   Assume that no three diagonal lines intersect at a interior point in a convex n-sided polygon. Thus the convex n-sided polygon is divided into several little regions by these diagonal lines. How many little regions are there in this convex n-sided polygon? (The selective examination for the 5th National Team of China)

Solution I   Suppose that among these little regions, there are n_3 triangles, n_4 convex quadrilateral, ..., and n_m convex m-sided polygon, then the total of little regions is

$$S = n_3 + n_4 + \cdots + n_m.$$

On the one hand, the total of vertices of all little regions is $3n_3 + 4n_4 + \cdots + mn_m$. On the other hand, the total of intersecting points of all diagonal lines in the convex n-sided polygon is $\binom{n}{4}$ and each intersecting point of the diagonal lines is the common vertex of four little regions (Since no three diagonal lines intersect at a interior point) and each vertex of the convex n-sided polygon is the common vertex of $n-2$ litter regions. Hence we obtain

$$3n_3 + 4n_4 + \cdots + mn_m = 4\binom{n}{4} + n(n-2). \qquad ①$$

Next, the total sum of interior angles of all little regions is

$$n_3 \cdot 180° + n_4 \cdot 360° + \cdots + n_m \cdot (n_m - 2) \cdot 180°.$$

On the other hand, the sum of interior angles of the n-sided polygon is $(n-2) \cdot 180°$ and the sum of interior angles at $\binom{n}{4}$ intersecting points of diagonal lines is $\binom{n}{4} \cdot 360°$. Hence

$$n_3 \cdot 180° + n_4 \cdot 360° + \cdots + n_m \cdot (m-2) \cdot 180°$$
$$= (n-2) \cdot 180° + \binom{n}{4} \cdot 360°,$$

i.e.

$$n_3 + 2n_4 + \cdots + (m-2)n_m = (n-2) + 2\binom{n}{4}. \qquad ②$$

With $[① - ②]/2$, we obtain

$$S = n_3 + n_4 + \cdots + n_m$$
$$= \binom{n}{4} + \frac{1}{2}(n-1)(n-2)$$
$$= \frac{1}{24}(n-1)(n-2)(n^2 - 3n + 12).$$

Solution Ⅱ   If we delete a diagonal line every time, then the number of regions decreases $a_i + 1$, where a_i equals the number of intersecting points of this diagonal line with the other diagonal lines. If all $\binom{n}{2} - n = \frac{1}{2}n(n-3)$ diagonal lines are deleted step by step, finally the number of the remaining regions is one. Therefore the required number of regions equals $1 + \sum_{i=1}^{n(n-3)/2} (a_i + 1)$. But $\sum_{i=1}^{n(n-3)/2} a_i$ equals $\binom{n}{4}$ (it is the number of the intersecting points of all diagonal lines in the convex n-sided polygon), so

$$S = 1 + \sum_{i=1}^{n(n-3)/2} (a_i + 1)$$

$$= 1 + \binom{n}{4} + \frac{1}{2}n(n-3)$$

$$= \frac{1}{24}(n-1)(n-2)(n^2 - 3n + 12).$$

Solution Ⅲ   Let the S denote the number of little regions. Adding the exterior region of the convex n-sided polygon, we obtain a planar graph with $F = S + 1$ regions and the number of its vertices equals $\binom{n}{4} + n$. Let E denote the number of its edges. Since the number of edges meeting each intersecting point of the diagonal lines in the convex n-sided polygon is 4, and the total is $4\binom{n}{4}$ and the number of edges on the diagonal meeting the each vertex of the convex n-sided polygon is $n - 3$, and the total is $n(n-3)$. But in the above counting, each edge is counted twice. Adding n edges of the convex n-sided polygon, we obtain

$$E = \frac{1}{2}\left[4\binom{n}{4} + n(n-3)\right] + n = 2\binom{n}{4} + \binom{n}{2}.$$

By Euler's formula $V + F - E = 2$, we obtain

$$S = F - 1 = E - V + 1$$

$$= 2\binom{n}{4} + \binom{n}{2} - \binom{n}{4} - n + 1$$

$$= \frac{1}{24}(n - 1)(n - 2)(n^2 - 3n + 12).$$

Solution IV Let the convex n-sided polygon be divided into a_n smaller regions by these diagonal lines, thus $a_3 = 1$, and $a_4 = 4$. If we add a point P_n and two line segments $P_n P_{n-1}$, $P_n P_1$ to the convex $(n - 1)$-sided polygon $P_1 P_2 \cdots P_{n-1}$, then we obtain a convex n-sided polygon and the polygon increases a triangle region $P_{n-1} P_n P_1$. If we add $n - 3$ diagonal lines $P_n P_2$, $P_n P_3$, $\cdots$, $P_n P_{n-2}$, then the number of increasing regions equals the sum of $n - 3$ and the number of the intersecting points on the $n - 3$ diagonal lines, i. e. $\binom{n}{4} - \binom{n-1}{4} +$ $(n - 3)$. Thus we obtain

$$a_n = a_{n-1} + 1 + \left(\binom{n}{4} - \binom{n-1}{4} \right) + (n - 3)$$

$$= a_{n-1} + \left(\binom{n}{4} - \binom{n-1}{4} \right) + (n - 2) \quad (n \geqslant 5).$$

For convenience, let $a_2 = 0$, $\binom{2}{4} = \binom{3}{4} = 0$, then for $n = 3$, 4, the above equality also holds. Therefore

$$a_n = a_2 + \sum_{k=3}^{n} (a_k - a_{k-1})$$

$$= 0 + \sum_{k=3}^{n} \left[\left(\binom{k}{4} - \binom{k-1}{4} \right) + (k - 2) \right]$$

$$= \binom{n}{4} + \frac{1}{2}(n - 1)(n - 2)$$

$$= \frac{1}{24}(n - 1)(n - 2)(n^2 - 3n + 12).$$

Example 5 Suppose that a_1, a_2, a_3, a_4, a_5 is a permutation of 1, 2, 3, 4, 5 satisfying that for any $i = 1, 2, 3, 4$, each permutation of a_1, a_2, $\ldots$, a_i is not a permutation of 1, 2, $\ldots$, i. Find the number of these permutations. (2000 Shanghai Mathematical Competition in China)

Solution I Obviously $a_1 \neq 1$.

(1) When $a_1 = 5$ and $a_2 a_3 a_4 a_5$ is a permutation of 1, 2, 3, 4, and the permutation $5 a_2 a_3 a_4 a_5$ satisfies the condition of the problem. In this case, the required number of the permutations is $4!$.

(2) When $a_1 = 4$, the number of permutations with the form $a_2 a_3 a_4 a_5$ is $4!$. But the permutations with the form $4 \times \times \times 5$ do not satisfy the condition of the problem and the number of these permutations is $3!$. Hence in this case, the required number of the permutations is $4! - 3!$.

(3) When $a_1 = 3$, the number of permutations with the form $a_2 a_3 a_4 a_5$ is $4!$. But the permutations with the forms $3 \times \times \times 5$ or $3 \times \times 54$ do not satisfy the condition of the problem. Hence in this case, the required number of the permutations is $4! - 3! - 2!$.

(4) When $a_1 = 2$, the number of permutations with the form $a_2 a_3 a_4 a_5$ is $4!$. But the permutations with the forms $215 \times \times$, 21453, $2 \times \times 54$ or $2 \times \times \times 5$ do not satisfy the condition of the problem. Hence in this case, the required number of the permutations is $4! - 2! - 1 - 2! - 3!$.

Summarizing what we have described above, the required number of the permutations equals

$$4! + (4! - 3!) + (4! - 3! - 2!) + (4! - 2! - 1 - 2! - 3!)$$
$$= 24 + 18 + 16 + 13 = 71.$$

Solution II Let the set of the permutations of 1, 2, 3, 4, 5 be S and $A_i = \{(a_1, a_2, a_3, a_4, a_5) \mid (a_1, a_2, a_3, a_4, a_5) \in S$ and $(a_1, a_2, \ldots, a_i)$ is a permutation of 1, 2, $\ldots$, i $(i = 1, 2, 3, 4)\}$. Thus the required number of the permutations is $|\overline{A_1} \cap \overline{A_2} \cap \overline{A_3} \cap \overline{A_4}|$. We obtain that

$|S| = 5!$, $|A_1| = 4!$, $|A_2| = 2! \cdot 3!$, $|A_3| = 3! \cdot 2!$,
$|A_4| = 4!$, $|A_1 \cap A_2| = 3!$, $|A_1 \cap A_3| = 2! \cdot 2!$,
$|A_1 \cap A_4| = 3!$, $|A_2 \cap A_3| = 2! \cdot 2!$, $|A_2 \cap A_4| = 2! \cdot 2!$,
$|A_3 \cap A_4| = 3!$, $|A_1 \cap A_2 \cap A_3| = 2!$,
$|A_1 \cap A_2 \cap A_4| = 2!$, $|A_1 \cap A_3 \cap A_4| = 2!$,
$|A_2 \cap A_3 \cap A_4| = 2!$, $|A_1 \cap A_2 \cap A_3 \cap A_4| = 1$.

Applying the inclusion – exclusion principle, we obtain that the required number of the permutations equals

$$|\overline{A_1} \cap \overline{A_2} \cap \overline{A_3} \cap \overline{A_4}|$$

$$= |S| - \sum_{i=1}^{4} |A_i| + \sum_{1 \leqslant i < j \leqslant 4} |A_i \cap A_j| - \sum_{1 \leqslant i < j < k \leqslant 4} |A_i \cap A_j \cap A_k|$$
$$+ |A_1 \cap A_2 \cap A_3 \cap A_4|$$

$$= 5! - (4! + 2! \cdot 3! + 3! \cdot 2! + 4!) + (3! + 2! \cdot 2! + 3!$$
$$+ 2! \cdot 2! + 2! \cdot 2! + 3!) - (2! + 2! + 2! + 2!) + 1$$

$$= 120 - 72 + 30 - 8 + 1 = 71.$$

Remark The reason why the equality $|A_1 \cap A_3| = 2! \cdot 2!$ holds is that the permutations in the set $A_1 \cap A_3$ have the forms $1 \times \times 45$ and $1 \times \times 54$. The reason why the equality $|A_1 \cap A_4| = 3!$ holds is that the permutations in the set $A_1 \cap A_4$ have the form $1 \times \times \times 5$ and so on.

Example 6 Let $M(p \times 1994, 7p \times 1994)$ be a point in the rectangular coordinate plane, where p is a prime. Find the number of rectangular triangles satisfying the following conditions.

(1) The vertices of the triangle are all integral points and the point M is the vertex of right angle;

(2) The incenter of the triangle is the origin.

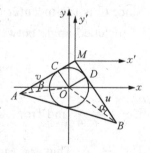

Analysis As in figure 13.1, the rectangular triangle MAB with the inradius r satisfies the conditions (1) and (2). Drawing two

Figure 13.1

perpendicular lines of MA and MB from O, and the feet of perpendicular lines are points C and D respectively. Thus $OC = OD = MC = MD = r = \dfrac{1}{\sqrt{2}}OM$ and $\text{Rt}\triangle MAB$ is determined uniquely by $BD = u$ and $AC = v$. We could find the number of required rectangular triangles if we establish the indefinite equation of u and v with conditions (1) and (2) and find the number of positive integral solutions of this indefinite equation. Therefore if we establish a corresponding relation, an unfamiliar problem may be translated into a familiar and easy problem.

Solution   To facilitate counting, the origin O is parallel translated to the point M, and we establish a new coordinate system $x'My'$. In the new coordinate system, the coordinates of points M and incenter O are $(0, 0)$ and $(-p \times 1994, -7p \times 1994)$ respectively.

Thus

$$| OM | = \sqrt{(p \times 1994)^2 + (7p \times 1994)} = p \times 1994 \times 5\sqrt{2}.$$

Suppose that $\triangle MAB$ with the inradius r satisfies the conditions (1) and (2). Drawing two perpendicular lines of MA and MB from O, and the feet of perpendicular lines are points C and D respectively. Assume that the slope of AM is k, thus we get the slope of OM as follows:

$$k' = \frac{7p \times 1994}{p \times 1994} = 7.$$

Since O is the incenter of $\triangle MAB$, and $\angle OMA = 45°$, by the formula of included angle between two straight lines, we obtain

$$1 = \tan 45° = \frac{k' - k}{1 - k'k} = \frac{7 - k}{1 - 7k},$$

i.e. $k = \dfrac{3}{4}$ and from $MB \perp MA$, we obtain that the slope of MB is $k_1 = -\dfrac{4}{3}$. Thus we suppose that the coordinates of A and B are

$A(-4t, -3t)$, and $B(3t', -4t')$. Since $(3, 4) = 1$ and A, B are integral points, then t and t' are two positive integers. It follows that $MA = 5t$, $MB = 5t'$ and

$$r = MC = MD = MO\cos 45° = p \times 1994 \times 5.$$

Denote $BD = u$, and $AC = v$, then

$$u = MB - MD = 5(t' - p \times 1994),$$

and

$$v = MA - MC = 5(t - p \times 1994).$$

Denote $\angle OBD = \alpha$, and $\angle OAC = \beta$, then

$$\alpha + \beta = \frac{1}{2}(\angle MAB + \angle MBA) = 45°,$$

$\tan \alpha = \dfrac{r}{u}$, $\tan \beta = \dfrac{r}{v}$ and

$$\frac{r}{u} = \tan \alpha = \tan(45° - \beta) = \frac{1 - \tan \beta}{1 + \tan \beta} = \frac{v - r}{v + r}.$$

Hence $u = \dfrac{r(v + r)}{v - r}$. Denote $v - r = m$, $\dfrac{2r^2}{m} = n$, then $u = \dfrac{r(m + 2r)}{m} = r + n$. Hence m and n are all multiples of 5. Denote $m_0 = \dfrac{m}{5}$, $n_0 = \dfrac{n}{5}$, then m_0, n_0 are two positive integers and

$$m_0 n_0 = \frac{m}{5} \cdot \frac{n}{5} = \frac{2r^2}{5^2} = \frac{2}{5^2}(p \times 1994 \times 5)^2 = 2^3 \times 977^2 \times p^2. \quad ①$$

Since each pair of the positive integers (m_0, n_0) corresponds to only one pair of positive integers $(u, v) = (n + r, m + r)$, we just obtain a triangle MAB satisfying the conditions of the problem. Hence the number S of triangles satisfying the conditions of the problem equals the number of the positive integer solutions (m_0, n_0) of equation ①.

(1) If $p \neq 2$, $p \neq 997$, then all solutions of equation $m_0 \cdot n_0 =$

$2^3 \times 997^2 \times p^2$ are

$$\begin{cases} m_0 = 2^i \times 997^j \times p^k, \\ n_0 = 2^{3-i} \times 997^{2-j} \times p^{2-k}, \end{cases}$$

where $i = 0, 1, 2, 3, j = 0, 1, 2, k = 0, 1, 2$.
Thus the number of the positive integer solutions (m_0, n_0) of equation
① equals $4 \times 3 \times 3 = 36$.

(2) If $p = 997$, then all solutions of equation $m_0 \cdot n_0 = 2^3 \times 997^4$
are

$$\begin{cases} m_0 = 2^i \times 997^j, \\ n_0 = 2^{3-i} \times 997^{4-j}, \end{cases}$$

where $i = 0, 1, 2, 3, j = 0, 1, 2, 3, 4$.
Thus the number of the positive integer solutions (m_0, n_0) of equation
① equals $4 \times 5 = 20$.

(3) If $p = 2$, then all solutions of equation $m_0 \cdot n_0 = 2^5 \times 997^2$ are

$$\begin{cases} m_0 = 2^i \times 997^j, \\ n_0 = 2^{5-i} \times 997^{2-j}, \end{cases}$$

where $i = 0, 1, 2, 3, 4, 5, j = 0, 1, 2$.
Thus the number of the positive integer solutions (m_0, n_0) of equation
① equals $6 \times 3 = 18$.

Summarizing what we have described above, we obtain that the number of rectangular triangles satisfying conditions (1) and (2) is

$$S = \begin{cases} 36 \ (p \neq 2, \ p \neq 997), \\ 20 \ (p = 997), \\ 18 \ (p = 2). \end{cases}$$

Solution II With the signs in solution I, we obtain that $MA = 5t$, $MB = 5t'$ (t, t' are positive integers) and $r = p \times 1994 \times 5$. Denote $r_0 = \dfrac{r}{5} = 2 \times 997 \times p$, thus

$$AB = \sqrt{MA^2 + MB^2} = 5\sqrt{t^2 + t'^2}.$$

Applying the conclusions of the plane geometry, we know that

$$AB = MA + MB - 2r = 5t + 5t' - 10r_0,$$

i. e.

$$5\sqrt{t^2 + t'^2} = 5(t + t' - 2r_0),$$

i. e.

$$\sqrt{t^2 + t'^2} = t + t' - 2r_0.$$

Squaring and rearranging the equality, we obtain

$$(t - 2r_0)(t' - 2r_0) = 2r_0^2 = 2^3 \times 997^2 \times p^2.$$

Denote $m_0 = t - 2r_0$, and $n_0 = t' - 2r_0$, then

$$m_0 n_0 = 2r_0^2 = 2^3 \times 997^2 \times p^2. \qquad \textcircled{2}$$

Therefore the number S of triangles satisfying conditions of the problem equals the number of the positive integer solutions (m_0, n_0) of the infinite equation ②. As the solution I , we find

$$S = \begin{cases} 36 \ (p \neq 2, \ p \neq 997), \\ 20 \ (p = 997), \\ 18 \ (p = 2). \end{cases}$$

Exercise 13

1 Each pair of the vertices of a cube determines a straight line. (1) Among these straight lines, how many pairs of straight lines are not in the same plane are there? (2) How many pairs of straight lines are mutually perpendicular in part (1)?

2 An equilateral triangle ABC of side length n is divided into n^2 equilateral triangles of side length 1 by lines parallel to its sides. (1) How many equilateral triangles are there? (2) How many rhombuses are there in this figure?

3 In a tournament each player played exactly one game against each of the other players. In each game the winner was awarded 2 points, the loser got 0 point, and each of two players earned 1 point if the game was a tie. After the completion of the tournament, it was found that exactly half of the points earned by each player were earned in the games against the ten players with the least number of points. (In particular, each of the ten lowest scoring players earned half of her/his points in the games against the other nine of the ten). How many players are there? (3nd American Invitational Mathematical Examination in 1985)

4 There are 18 points in the plane, no three of which are collinear. Every two points are connected with a line segment and every line segment is colored with red or blue. Assume that the number of red line segments meeting the point A is an odd and the numbers of red line segments meeting each of other 17 points are different. Find the number of triangles whose three sides are red and the number of triangles whose two sides are red and the third side is blue and whose vertices are the given 18 points.

5 The permutation $(n_1, n_2, n_3, n_4, n_5, n_6)$ of 1, 2, 3, 4, 5, 6 has the following property for any $i = 1, 2, 3, 4, 5$, each permutation $n_1, n_2, \ldots, n_i$ is not the permutation of 1, 2, $\ldots$, i. Find the number of these permutations. (1991 British Mathematical Olympiad)

6 The $4 \times 4 \times 4$ cube consists of 64 unit cubes. Then 16 unit cubes are selected and colored red such that each $1 \times 1 \times 4$ rectangular solids consisting of 4 unit cubes in this $4 \times 4 \times 4$ cube contains exactly a red unit cube. How many ways to select 16 unit cubes are there? Justify your answer. (14th CMO)

In mathematical competition, we often prove that there exist some combinatorial constructions with some given properties. To solve these problems, we may use the following methods.

(1) Reduction to absurdity and the extreme principle (Examples 1, 3, and 5 – 7, in Chapter 10);

(2) The pigeonhole principle, mean value principle and overlap principle of figures (Examples 1 – 3, and 5 – 11 in Chapter 2);

(3) Counting method (Examples 3, and 13 – 15 in Chapter 6, Examples 7 – 8 in Chapter 8);

(4) Colored method and Evaluation method (Examples 1, 5 – 7 in Chapter 9);

(5) Mathematical inductive method (Example 8 in Chapter 2, Example 2 in Chapter 5, Example 7 in Chapter 12);

(6) Combinatorial analytical method (Examples 3 – 6 in Chapter 5);

(7) Construction method (Examples 1 – 7 in Chapter 12).

Sometimes, we not only prove that some combinatorial constructions satisfying the given conditions are existential but also determine that the number of these combinatorial constructions is changed within a definite range. Thus we should prove some inequalities concerning the number of the combinatorial constructions. Using the above methods (1)-(6), we can prove some inequalities in combinatorial problems. Examples 12 – 13 in Chapter 6 and Example 4, and 6 in Chapter 7 are examples about proving the inequalities in the combinatorial problems. Furthermore, we will show some other examples of the methods to prove the inequalities in the combinatorial problems in the following depiction.

Example 1 Suppose that we put 9 small squares with area $\frac{1}{5}$ in the planar figure F with area 1 at random. Prove that there exist at least two small squares such that the area of their intersection is not less than $\frac{1}{45}$.

Proof Suppose the 9 small squares are A_1, A_2, $\ldots$, A_9 and denote the area of A_i by $|A_i|$ $(i = 1, 2, \ldots, 9)$. Since $\sum\limits_{i=1}^{9} |A_i| = \frac{9}{5} > 1$, there exist two small squares overlapping. If for any $1 \leqslant i < j \leqslant 9$, $|A_i \cap A_j| < \frac{1}{45}$, then the area of part which is covered by A_1, A_2, $\ldots$, A_9 equals

$$|A_1 \cup A_2 \cup \cdots \cup A_9|$$

$$\geqslant \sum_{i=1}^{9} |A_i| - \sum_{1 \leqslant i < j \leqslant 9} |A_i \cap A_j|$$

$$> 9 \times \frac{1}{5} - \binom{9}{2} \times \frac{1}{45} = 1.$$

But A_1, A_2, $\ldots$, A_9 are all contained in the planar figure F with the area 1, hence

$$|A_1 \cup A_2 \cup \cdots \cup A_9| \leqslant 1,$$

and it is a contradiction. Therefore there exist two small squares A_i and $A_j (1 \leqslant i < j \leqslant 9)$ such that $|A_i \cap A_j| \geqslant \frac{1}{45}$.

Remark In Example 1, we prove it by contradiction and the following overlap principle of figures.

Overlap principle of figures Suppose that we put n planar figures A_1, A_2, $\ldots$, A_n with areas S_1, S_2, $\ldots$, S_n respectively in a planar figure F with the area S.

(1) If $S_1 + S_2 + \cdots + S_n > S$ then there are two planar figures A_i and $A_j (1 \leqslant i < j \leqslant n)$ with common interior points.

(2) If $S_1 + S_2 + \cdots + S_n < S$, then there exists at least one point $P \in F$ which does not belong to each of the planar figures A_1, A_2, ..., A_n.

Example 2   Suppose that in a certain society, each pair of persons can be classified as either amicable or hostile. We shall say that each member of an amicable pair is a friend of the other, and each member of a hostile pair is a foe of the other. Suppose that the society has n persons and q amicable pairs, and that for every set of three persons, at least one pair is hostile. Prove that there is at least one member of the society such that the number of amicable pairs in the set of his (or her) foes is less than or equal to $q \left(1 - \dfrac{4q}{n^2} \right)$ (24th United States of America Mathematical Olympiad in 1995)

Proof I   Let n persons be a_1, a_2, ..., a_n and the set of amicable pairs be S among these persons. Let the number of friends of a_i be $d_i (i = 1, 2, \ldots, n)$, thus $\sum\limits_{i=1}^{n} d_i = 2 \mid S \mid = 2q$.

We construct an $n \times n$ table such that the number in the ith row and the jth column is

$$x_{ij} = \begin{cases} 1, \text{ if } i \neq j \text{ and } (a_i, a_j) \text{ is an amicable pair,} \\ 0, \text{ if } i = j \text{ or } i \neq j \text{ and } (a_i, a_j) \text{ is a hostile pair,} \end{cases}$$

where i, $j = 1, 2, \ldots, n$.
Thus

$$d_i = \sum_{j=1}^{n} x_{ij} = \sum_{j=1}^{n} x_{ji} (i = 1, 2, \ldots, n),$$

and $\sum\limits_{j=1}^{n} x_{ij} d_j$ expresses the number of the amicable pairs satisfying that among these pairs, there is a person who is a_i or the friend of a_i (Since for every set of three persons, at least one pair is hostile, the counting in this expression is not repeated). Hence $\mid S \mid - \sum\limits_{j=1}^{n} x_{ij} d_j$ is the number of amicable pairs among the foes of a_i. By Cauchy's inequality, we obtain

$$\sum_{i=1}^{n} \sum_{j=1}^{n} x_{ij} d_j = \sum_{j=1}^{n} \left(\sum_{i=1}^{n} x_{ij} \right) d_j$$

$$= \sum_{j=1}^{n} d_j^2 \geqslant \left(\sum_{j=1}^{n} d_j \right)^2 / \sum_{j=1}^{n} 1^2$$

$$= \frac{4q^2}{n}.$$

Thus

$$\frac{1}{n} \sum_{i=1}^{n} \sum_{j=1}^{n} x_{ij} d_j \geqslant \frac{4q^2}{n^2}.$$

By the mean value principle, we know that there exists i_0 ($1 \leqslant i_0 \leqslant n$) such that $\displaystyle\sum_{j=1}^{n} x_{i_0 j} d_j \geqslant \frac{4q^2}{n^2}$, i.e.

$$|S| - \sum_{j=1}^{n} x_{i_0 j} d_j \leqslant q - \frac{4q^2}{n^2} = q \left(1 - \frac{4q}{n^2} \right).$$

So the number of amicable pairs in the set of his (or her) foes is less than or equal to $q \left(1 - \dfrac{4q}{n^2} \right)$.

Proof II Denote n persons by n points A_1, A_2, $\ldots$, A_n. If two persons are amicable, then two corresponding points are connected by a red line segment. If two persons are hostile, then two corresponding points are connected by a blue line segment, thus we obtain a 2-colored complete graph K_n. Let d_i denote the number of red edges meeting the point A_i, i.e. the A_i has d_i friends, thus the number of blue edges meeting the point A_i is $n - 1 - d_i$, i.e. A_i has $n - 1 - d_i$ foes ($i = 1$, 2, $\ldots$, n). Hence

$$\sum_{i=1}^{n} d_i = 2q. \tag{①}$$

Suppose that the number of triangles with two blue sides and a red side is A, and the number of triangles with two red sides and a blue side is B. According to the given conditions, there is not a triangle with three red sides, hence

$$B = \sum_{i=1}^{n} \binom{d_i}{2} \quad \text{(When } d_i = 0 \text{ or } 1, \binom{d_i}{2} = \frac{1}{2} d_i (d_i - 1) = 0 \text{)}. \quad \text{②}$$

If a vertex of an angle is A_i $(i = 1, 2, \ldots, n)$ and colors of its two sides meeting A_i are different, then this angle is called a different color angle. Thus total of the different color angles equals

$$2(A + B) = \sum_{i=1}^{n} d_i (n - 1 - d_i). \quad \text{③}$$

With ①, ② and ③ and by Cauchy's inequality, we obtain

$$A = \frac{1}{2} \sum_{i=1}^{n} d_i (n - 1 - d_i) - B$$

$$= \frac{1}{2} \sum_{i=1}^{n} d_i (n - 1 - d_i) - \frac{1}{2} \sum_{i=1}^{n} d_i (d_i - 1)$$

$$= \frac{n}{2} \sum_{i=1}^{n} d_i - \sum_{i=1}^{n} d_i^2$$

$$\leqslant \frac{n}{2} \sum_{i=1}^{n} d_i - \frac{1}{n} \left(\sum_{i=1}^{n} d_i \right)^2$$

$$= nq - \frac{4q^2}{n} = nq \left(1 - \frac{4q^2}{n^2} \right). \quad \text{④}$$

Suppose that (B_j, B_k) is an amicable pair among the foes of A_i. thus the triangle $A_i B_j B_k$ has two blue sides and a red side. We use x_i denote the number of triangles which has two blue sides $A_i B_j$, $A_i B_k$ and a red side $B_j B_k$, thus there are x_i amicable pairs among the foes of A_i and $\sum_{i=1}^{n} x_i = A$. From ④ we obtain

$$\frac{1}{n} \sum_{i=1}^{n} x_i = \frac{1}{n} A \leqslant q \left(1 - \frac{4q}{n^2} \right).$$

By the mean value principle, we know that there exists a x_i satisfying $x_i \leqslant q \left(1 - \frac{4q}{n^2} \right)$, i.e. foes of A_i include $q \left(1 - \frac{4q}{n^2} \right)$ or fewer amicable pairs.

Example 3   Suppose that each integral point in the plane is colored with one of the three colors — red, blue and green and every color is used. Prove that there exists a rectangular triangle such that three vertices of this triangle are colored with different colors. (30th Russian Mathematical Olympiad)

Proof   When all integral points in vertical lines have the same colors, we take any one integral point A (Assume that A is colored red). Draw two straight lines l_1 and l_2 from point A such that the angles between the vertical line and the two straight lines equal $45°$, and we take a blue integral point B lying on l_1 and a green integral point C lying on l_2 respectively (Since there exist vertical lines of blue and green points and these vertical lines and $l_1(l_2)$ intersect at integral points, we can take the points B and C). Thus $\triangle ABC$ satisfies the requirement of the problem. Analogously, when all integral points in each horizontal line have the same color, the conclusion is also true.

If there exists a vertical line l such that all points of l are colored with two different colors (red and blue), then we take any green point C outside l and draw one perpendicular line of l from C. Thus the perpendicular line intersects l at an integral point A. (Assume that A is colored red). Take a blue integral point B on l, thus $\triangle ABC$ satisfies the condition of the problem.

If there exists a vertical line l such that all points of this vertical line are colored with three different colors (red, blue and green), then we take any horizontal line m such that the points of m are at least colored with two different colors. Assume the intersecting point A of l and m is colored red, we take a blue integral point B of m and a green integral point C on l of l, thus $\triangle ABC$ satisfies the condition of the problem.

Example 4   Let S be a set with 2002 elements, and let N be an integer with $0 \leqslant N \leqslant 2^{2002}$. Prove that it is possible to color every subset of S either blue or red so that the following conditions hold:

(a) The union of any two red subsets is red;

(b) The union of any two blue subsets is blue;

(c) There are exactly N red subsets. (31st United States of America Mathematical Olympiad)

Proof We prove that this can be done for any n-element set, where n is a positive integer, $S_n = \{1, 2, \ldots, n\}$ and for any integer N_n with $0 \leqslant N_n \leqslant 2^n$.

We induct on n. The base case $n = 1$ is trivial. Assume that we can color the subsets of $S_n = \{1, 2, \ldots, n\}$ in the desired manner, for any integer N_n with $0 \leqslant N_n \leqslant 2^n$. We show that there is a desired coloring for $S_{n+1} = \{1, 2, \ldots, n+1\}$ and the integer N_{n+1} with $0 \leqslant N_{n+1} \leqslant 2^{n+1}$. We consider the following cases.

(1) $0 \leqslant N_{n+1} \leqslant 2^n$. Applying the induction hypothesis to S_n and $N_n = N_{n+1}$, we get a coloring of all subsets of S_n satisfying conditions (a), (b), (c). All uncolored subsets of S_{n+1} contain the element $n + 1$, we color all of them blue. It is not hard to see this coloring of all the subsets of S_{n+1} satisfying conditions (a), (b), (c).

(2) $2^n + 1 \leqslant N_{n+1} \leqslant 2^{n+1}$. Without loss of generality, let $N_{n+1} = 2^n + k$ ($k = 1, 2, \ldots, 2^n$). By case (1), we know that there exists a coloring of the subsets of S_n satisfying conditions (a) and (b) and having k red subsets. All uncolored 2^n subsets of S_{n+1} contain the element $n + 1$, and we color all of them red. Then, the conditions (a) and (b) are satisfied and the number of red subsets equals $2^n + k = N_{n+1}$, i. e. the condition (c) is satisfied. Thus our induction is completed.

Example 5 Assume that S is a finite point set whose elements are colored with red or blue and $A_1, A_2, \ldots, A_{68}$ are the 5-element subsets of S satisfying the following conditions:

(1) There is at least one red point in each of $A_1, A_2, \ldots, A_{68}$;

(2) For any three points in S, there exists exactly one subset A_i containing the three points. Is there a subset A_i containing four or five red points? Prove your conclusion.

Solution I Assume that S contains n elements, then S has $\binom{n}{3}$

3-element subsets and each 3-element set is contained in exactly one of $A_1, A_2, \ldots, A_{68}$. On the other hand, each A_i has $\binom{5}{2}$ 3-element subsets, and the total of 3-element subsets of $A_1, A_2, \ldots, A_{68}$ is $68\binom{5}{2}$. Thus we know that $\binom{n}{3} = 68\binom{5}{2}$, i. e. $n = 17$.

Assume that there are r red points in S, and each of $A_1, A_2, \ldots, A_{68}$ contains at most three red points. A 3-element subset B of S is called a red 3-element subset, if B contains exactly one red point and two blue points. Thus the total of red 3-element subsets of S is $\binom{r}{1}$ $\binom{17-r}{2}$. On the other hand, for any three red points, there exists exactly one 5-element subset A_i containing the three red points, hence there are $\binom{r}{3}$ 5-element subsets with three red points and each such 5-element subsets contains $\binom{3}{1}\binom{2}{2} = 3$ red 3-element subsets. Next, each of the remaining $68 - \binom{r}{3}$ 5-element subsets contains exactly one or two red points and each such 5-element subsets contains $\binom{1}{1}\binom{4}{2} = 6$ or $\binom{2}{1}\binom{3}{2} = 6$ red 3-element subsets, hence the total of red 3-element subsets is $3\binom{r}{3} + 6\left(68 - \binom{r}{3}\right)$. Thus we obtain that

$$3\binom{r}{3} + 6\left(68 - \binom{r}{3}\right) = \binom{r}{1}\binom{17-r}{2}.$$

Rearranging the equality above, we obtain $r^3 - 18r^2 + 137r = 408$, thus

$$3 \equiv r^3 - 3r^2 + 2r \equiv r(r-1)(r-2) \pmod{5}.$$

But when $r \equiv 0, 1, 2, 3, 4 \pmod 5$, the equality above does not hold. It leads to a contradiction. Therefore we have proved that there is a subset A_i containing four or five red points.

Solution II   From solution I, we know $|S| = 17$. Assume that the red points in S are $P_1, P_2, \ldots, P_r$, and each of $A_1, A_2, \ldots, A_{68}$ contains at most three red points. Let k_i denote the number of 5-element subset containing the red point P_i $(1 \leqslant i \leqslant r)$. On the one hand, the number of 3-element subsets of S containing the red point P_i equals $\binom{16}{2} = 120$. On the other hand, there are $\binom{4}{2} = 6$ 3-element subsets containing the red point P_i in each of 5-element subsets containing the red point P_i. Hence $6k_i = 120$, i.e. $k_i = 20$ $(1 \leqslant i \leqslant r)$. Let k_{ij} denote the number of 5-element subset containing the red points P_i and P_j $(1 \leqslant i < j \leqslant r)$. On one hand, the number of 3-element subsets of S containing the red points P_i and P_j equals $\binom{15}{1} = 15$. On the other hand, there are $\binom{3}{1} = 3$ 3-element subsets containing the red points P_i and P_j in each of 5-element subsets containing the red points P_i and P_j. Thus $3k_{ij} = 15$, i.e. $k_{ij} = 5$ $(1 \leqslant i < j \leqslant r)$. Let k_{ijk} denote the number of 5-element subset containing the red points P_i, P_j and P_k $(1 \leqslant i < j < k \leqslant r)$. By the given conditions, we know $k_{ijk} = 1$, and with the hypothesis, any one of $A_1, A_2, \ldots, A_{68}$ does not contain 4 or 5 red points in $\{P_1, P_2, \ldots, P_r\}$. On one hand, from the given condition (1), we know that each of $A_1, A_2, \ldots, A_{68}$ contains at least one red point. On the other hand, by the inclusion – exclusion principle, we know that among $A_1, A_2, \ldots, A_{68}$ the number of sets containing at least one red point equals

$$\sum_{i=1}^{r} k_i - \sum_{1 \leqslant i < j \leqslant r} k_{ij} + \sum_{1 \leqslant i < j < k \leqslant r} k_{ijk}$$

$$= 20r - 5\binom{r}{2} + \binom{r}{3}$$

$$= \frac{1}{6}(r^3 - 18r^2 + 137r).$$

Thus

$$\frac{1}{6}(r^3 - 18r^2 + 137r) = 68,$$

i. e.

$$r^3 - 18r^2 + 137r = 408.$$

The the remaining steps are the same as solution I .

Example 6   Let A be an n-element set and A_1, A_2, $\ldots$, A_m be the subsets of A satisfying that any one of A_1, A_2, $\ldots$, A_m is not a subset of any other. Prove that

$$(1) \sum_{i=1}^{m} \frac{1}{\binom{n}{|A_i|}} \leqslant 1; \qquad (2) \sum_{i=1}^{m} \binom{n}{|A_i|} \geqslant m^2,$$

where $|A_i|$ denote the number of elements of the set $A_i (1 \leqslant i \leqslant m)$, $\binom{n}{|A_i|} = \dfrac{n!}{|A_i|!(n-|A_i|)!}$. (China Mathematical Competition in 1993)

Proof   (1) The inequality in (1) is equivalent to the following inequality:

$$\sum_{i=1}^{m} |A_i|!(n-|A_i|)! \leqslant n!. \qquad \qquad ①$$

On one hand, the number of all permutations of n elements of A equals $n!$. On the other hand, for each subset A_i, we construct all permutations of n elements of A as follows:

$$x_1 x_2 \cdots x_{|A_i|} y_1 y_2 \cdots y_{n-|A_i|}, \qquad \qquad ②$$

where $x_1 x_2 \cdots x_{|A_i|}$ is a all permutation of elements of A_i and $y_1 y_2 \cdots y_{n-|A_i|}$ is a all permutation of elements of $\overline{A_i}$ ($\overline{A_i}$ is the complement of A_i in A). Thus the number of permutations with the form as ② equals $|A_i|! \ (n-|A_i|)!$.

Next, we prove that if $i \neq j$, then the all permutations corresponding to A_i and A_j are different. In fact, assume that any all permutation corresponding to A_j is

$$x'_1 x'_2 \cdots x'_{|A_j|} y'_1 y'_2 \cdots y'_{n-|A_j|}. \tag{3}$$

Assume when $i \neq j$, ② and ③ are the same, then $x_1 = x'_1$, $x_2 = x'_2$, $\ldots$, $x_{|A_i|} = x'_{|A_i|}$, i.e. $A_i \subseteq A_j$ if $|A_i| \leqslant |A_j|$ or $x_1 = x'_1$, $x_2 = x'_2$, $\ldots$, $x_{|A_j|} = x'_{|A_j|}$, i.e. $A_j \subseteq A_i$ if $|A_j| < |A_i|$. It contradicts the given condition. Therefore we yield

$$\sum_{i=1}^{m} |A_i|! (n - |A_i|)! \leqslant n!.$$

Thus we deduce

$$\sum_{i=1}^{m} \frac{1}{\binom{n}{|A_i|}} \leqslant 1.$$

(2) With (1) and Cauchy's inequality, we obtain

$$\sum_{i=1}^{m} \binom{n}{|A_i|} \geqslant \left(\sum_{i=1}^{m} \binom{n}{|A_i|} \right) \left[\sum_{i=1}^{m} \frac{1}{\binom{n}{|A_i|}} \right] \geqslant m^2.$$

Remark Applying the inequality in (1), we can prove the Sperner Theorem as follows.

Sperner Theorem Let A be one n-element set, and $A_1, A_2, \ldots,$ A_m be the subsets of A satisfying that any one of $A_1, A_2, \ldots, A_m$ is not a subset of any other. Then the maximum of m is $\binom{n}{[n/2]}$.

Proof By the inequality in Example 6(1), we know

$$\sum_{i=1}^{m} \frac{1}{\binom{n}{|A_i|}} \leqslant 1.$$

Since the maximum in $\left\{ \binom{n}{1}, \binom{n}{2}, \ldots, \binom{n}{n} \right\}$ is $\binom{n}{[n/2]}$,

$$\frac{m}{\binom{n}{[n/2]}} \leqslant \sum_{i=1}^{m} \frac{1}{\binom{n}{|A_i|}} \leqslant 1,$$

i.e. $m \leqslant \binom{n}{[n/2]}$. On the other hand, the number of $[n/2]$-element

subsets of A equals $\binom{n}{[n/2]}$ and any one of these subsets is not a

subset of any other. Thus we have proved that the maximum of m

is $\binom{n}{[n/2]}$.

Example 7   In a competition, there are a contestants and b judges, where $b \geqslant 3$ is odd. Each judge rates each contestant as either "pass" or "fail". Suppose k is a number such that, for any two judges, their ratings coincide for at most k contestants. Prove that $\frac{k}{a} \geqslant \frac{b-1}{2b}$.

(39th IMO)

Proof   Let A_1, A_2, $\ldots$, A_a denote a contestants and B_1, B_2, $\ldots$, B_b denote b judges. If ratings of two judges B_i, $B_j (i \neq j)$ coincide with the ratings of the contestant A_m, thsen A_m, B_i, B_j form a triple $\{A_m; B_i, B_j\}$ and assume that the number of all triples is M.

On one hand, since for any two judges B_i, B_j, their ratings coincide with at most k contestants', there are at most k triples containing B_i, B_j, and there are exactly $\binom{b}{2}$ ways to select B_i, B_j. Hence

$$M \leqslant k\binom{b}{2}. \qquad\qquad ①$$

On the other hand, for any contestant A_m, assume that there are x judges whose ratings are "pass" for the contestant A_m and there are y judges whose ratings are "fair" for the contestant A_m. Thus $x + y = b$ and there are $\binom{x}{2} + \binom{y}{2}$ triples containing A_m, and there are a ways to select A_m, and $M = a\left(\binom{x}{2} + \binom{y}{2}\right)$. Since

$$\binom{x}{2} + \binom{y}{2} = \frac{x(x-1)}{2} + \frac{y(y-1)}{2}$$

$$= \frac{(x+y)^2 + (x-y)^2 - 2(x+y)}{4}$$

$$= \frac{b^2 - 2b + (x-y)^2}{4}$$

$$\geqslant \frac{b^2 - 2b}{4} = \frac{(b-1)^2 - 1}{4},$$

and $b \geqslant 3$ is odd,

$$\binom{x}{2} + \binom{y}{2} \geqslant \frac{(b-1)^2}{4} = \left(\frac{b-1}{2}\right)^2.$$

It follows that

$$M = a\left(\binom{x}{2} + \binom{y}{2}\right) \geqslant a\left(\frac{b-1}{2}\right)^2. \qquad ②$$

From ① and ②, we obtain

$$a\left(\frac{b-1}{2}\right)^2 \leqslant k \cdot \frac{b(b-1)}{2},$$

i. e. $\dfrac{k}{a} \geqslant \dfrac{b-1}{2b}$. This completes the proof.

Exercise 14

1   For which integers $n \geqslant 5$ is it possible to color the vertices of a regular n-sided polygon using at most 6 colors such that any 5 consecutive vertices have different colors? (2000 Austrian – Polish Mathematical Olympiad)

2   Every cell of a 100×100 board is colored with one of 4 colors

so that there exactly 25 cells of each color in every column and in every row. Prove that one can choose two columns and two rows so that the four cells where they intersect are colored in four different colors. (26th Russian Mathematical Olympiad in 2000)

3   There are 1001 cities in some country Ω and any two of them are joined by one unidirectional road. Every city has exactly 500 unidirectional roads out of this city and there are exactly 500 unidirectional roads into this city. Republic Γ containing 668 of the cities secedes from Ω. Prove that it is possible to travel between any two cities in this republic Γ without leaving the republic Γ. (30th Russian Mathematical Olympiad)

4   If the set $M = A_1 \cup A_2 \cup \cdots \cup A_n$ and $A_i \cap A_j = \varnothing$ $(1 \leqslant i < j \leqslant n)$, then the sets $A_1, A_2, \ldots, A_n$ are called the n-partition of M. Assume that $A_1, A_2, \ldots, A_n$ and $B_1, B_2, \ldots, B_n$ are two n-partitions of M and for any two disjoint sets A_i and B_j $(1 \leqslant i, j \leqslant n)$, the inequality $\mid A_i \cup B_j \mid \geqslant n$ holds. Prove that $\mid M \mid \geqslant \dfrac{n^2}{2}$.

5   A mathematical competition consists of two parts (first test and second test) and 28 questions in total. Each payer solved exactly 7 questions. Each pair of questions was solved by just two players. Show that one player solved either nil or at least 4 questions in the first test. (13th United States of America Mathematical Olympiad)

6   Assume that there are n points in space, no four points of which are coplanar and there are q line segments connecting these n points. Prove that there exist at least $\dfrac{4q}{3n}\left(q - \dfrac{n^2}{4}\right)$ triangles which consist of these line segments.

7   Let a space figure consist of n vertices and l lines connecting these vertices, with $n = q^2 + q + l$, $l \geqslant \dfrac{1}{2}q(q+1)^2 + 1$, $q \geqslant 2$, $q \in \mathbf{N}_+$. Suppose the figure satisfies the following conditions: any four vertices are non-coplanar, and every vertex is connected by at least one line, and there is a vertex which is connected by at least $q + 2$ lines. Prove that there exists a space quadrilateral in the figure, i. e. a quadrilateral with four vertices A, B, C, D and four lines AB, BC, CD, DA in the figure. (China Mathematical Competition in 2003)

    8   Let $X_n = \{1, 2, 3, \ldots, n^2 - n\}$. Prove that X_n can be partitioned into two disjoint nonempty subsets S_n and T_n such that each of S_n and T_n does not contains n elements a_1, a_2, $\ldots$, a_n $(a_1 < a_2 < \cdots < a_n)$ satisfying $a_k \leqslant \dfrac{a_{k-1} + a_{k+1}}{2}$ $(k = 2, 3, \ldots, k-1)$. (2008 CMO)

Chapter 15 Combinatorial Extremum Problems

The combinatorial extremum problems are important and interesting in the mathematical competitions. We describe this class of problems as follows.

Let $\mathfrak{R}$ be the set of some combinatorial constructions and $\mathfrak{S}$ be the subset of $\mathfrak{R}$ with the elements satisfying the given condition P. Every element A of $\mathfrak{R}$ corresponds to an unique real number $m = f(A)$. Our problem is that when $A \in \mathfrak{S}$, find the maximum or minimum of $f(A)$.

In the combinatorial extremum problems, the arguments usually are discrete combinatorial constructions such as positive integers, sets or graphs and so on and the relation of function often is not written in analytical expressions. Hence it is different to solve combinatorial extremum problems and algebraic extremum problems.

The basic steps of solving the combinatorial extremum problems list as follows.

(1) Explore the maximum (or minimum) m_0 of the required problem;

(2) Prove that for all $A \in \mathfrak{S}$, $m = f(A) \leqslant m_0 (\geqslant m_0)$;

(3) Construct a $A_0 \in \mathfrak{S}$ such that $f(A_0) = m_0$ holds. Thus we obtain that when $A \in \mathfrak{S}$, the maximum (or minimum) of $f(A)$ is $f(A_0) = m_0$.

For some combinatorial extremum problems, we may proceed with the following steps $(2)'$, $(3)'$ substituting the above steps (2), (3).

$(2)'$ Prove that any A of $\mathfrak{R}$ satisfying that $m = f(A) \leqslant m_0 (\geqslant m_0)$ belongs to $\mathfrak{S}$ (i.e. A satisfies the given condition P);

(3)$'$ When $m > m_0 (< m_0)$, we construct an $A_0 \in \Re$ such that $m = f(A_0) > m_0 (< m_0)$ and $A_0 \notin \Im$ (i. e. A_0 does not satisfy the given condition P).

When we solve the problem, the steps (1) and (2) (or (1) and (3)$'$) often are followed simultaneously, i. e. we explore and find the maximum (or minimum) m_0 of the required problem with analysis and proof or construction.

When $\Re$ (or $\Im$) is a finite set, the set of $m = f(A)$ is also finite. Hence there exists a combinatorial construction A_0 such that $f(A_0)$ is the maximum (or minimum) of $f(A)$. In this case, we assume that $f(A)$ take the maximum (or minimum) $f(A_0)$ at $A = A_0$, afterwards we often use the method of successive adjustment to find the necessary conditions which are satisfied by A_0. If A satisfying these necessary conditions is unique, then this A is the A_0 we want and the corresponding $f(A_0)$ is the required maximum (or minimum). If there are only few A satisfying these necessary conditions, we denote A_1, $A_2, \ldots, A_k$, then we compute the corresponding values $f(A_1)$, $f(A_2), \ldots, f(A_k)$, thus the required maximum (or minimum) of $f(A)$ equals $\underset{1 \leqslant i \leqslant k}{\text{Max}} \{f(A_i)\}$ (or $\underset{1 \leqslant i \leqslant k}{\text{Min}} \{f(A_i)\}$).

With the above analysis, we know that to find the solution of a combinatorial extremum problem, the proof and construction are necessary. We could consult the ways introduced in Chapters 10 - 12 to proceed with proof and construction. The main ways of exploring the extremum as follows.

(1) The method of estimating value. The usual ways of estimating extremum list as follows: constructing especial examples, considering especial cases, carrying out the whole estimate, considering extreme cases, and starting with opposite cases and so on. (Example 4 in Chapter 2; Example 5 in Chapter 5 and Problem 5 in Exercise 11);

(2) The method of combinatorial analysis. (Examples 2 and 7 in Chapter 2; Examples 2 and 4 in Chapter 9; Example 8 in Chapter 10);

(3) Counting method (Example 4 in Chapter 6; Problem 2 in

Exercise 7);

    (4) Adjustment method (Examples 2 and 3 in Chapter 11);

    (5) Mathematical induction method (Problem 4 in Exercise 11).

Example 1   Let $M = \{1, 2, \ldots, 1995\}$ and the set $A \subseteq M$ satisfy the following condition: for any $x \in A$, and $15x \notin A$. Find the maximum of $|A|$. (China Mathematical Competition in 1995)

Solution   Since $x \in A$, and $15x \notin A$, if $15x > 1995$, i. e. $x > 133$, then $15x \notin A$, then we know $\{134, 135, \ldots, 1995\} \subset A$. Next if $x \in A$, and $15x \notin A$, when $15x < 134$, i. e. $x < 9$, thus we know $\{1, 2, \ldots, 8\} \subset A$. Therefore the set

$$A = \{1, 2, \ldots, 8\} \cup \{134, 135, \ldots, 1995\}$$

satisfies the condition: for any $x \in A$, and $15x \notin A$ and

$$|A| = 1995 - (133 - 8) = 1870.$$

On the other hand, let A be any subset of M and A satisfying the given condition. Since for any $x \in \{9, 10, 11, \ldots, 133\}$, there is at most one of x and $15x$ belonging to A,

$$|A| \leqslant 1995 - (133 - 8) = 1870.$$

Summarizing what we have described above, we get that the maximum of $|A|$ equals 1870.

Example 2   Assume that a rectangle whose side lengths are positive integers m, n is perfectly divided into several squares whose side lengths are positive integers and sides are parallel to the corresponding sides of the rectangle. Find the minimum of the sum of sides of all squares. (China Mathematical Competition in 2001)

Analysis   Without loss of generality, assume that $m \geqslant n$, and we construct a special partition as follows. Firstly, we divide the $m \times n$ rectangle into squares with side length n and an $n \times r_1$ rectangle ($0 < r_1 < n$) Next, we divide the $n \times r_1$ rectangle into squares with side length r_1 and an $r_1 \times r_2 (0 < r_2 < r_1)$ rectangle. We continue this process till the rectangle is perfectly divided into little squares without remainder. Obviously, the process of dividing above is equivalent to the division algorithm of the two positive integers m and n.

Solution　Without loss of generality, assume that $m \geqslant n$ and $f(m, n)$ denote the minimum of sum of sides of all squares. With the division algorithm of the two positive integers m and n, we know that there exist several positive integers $q_1, q_2, \ldots, q_{k+1}$ and $r_1, r_2, \ldots, r_k$ such that

$$m = q_1 n + r_1 \, (0 < r_1 < n),$$

$$n = q_2 r_1 + r_2 \, (0 < r_2 < r_1),$$

$$r_1 = q_3 r_2 + r_3 \, (0 < r_3 < r_2),$$

$$\cdots$$

$$r_{k-2} = q_k r_{k-1} + r_k \, (0 < r_k < r_{k-1}),$$

$$r_{k-1} = q_{k+1} r_k.$$

Thus firstly we divide the $m \times n$ rectangle into $q_1 \, n \times n$ squares and an $n \times r_1$ rectangle. Next, we divide the $n \times r_1$ rectangle into $q_2 \, r_1 \times r_1$ squares and an $r_1 \times r_2$ rectangle $\ldots\ldots$ For the kth step, we divide the $r_{k-1} \times r_{k-2}$ rectangle into $q_k \, r_{k-1} \times r_{k-1}$ squares and an $r_k \times r_{k-1}$ rectangle. Finally, for the $(k+1)^{\text{th}}$ step, we divide the $r_k \times r_{k-1}$ rectangle into $q_{k+1} \, r_k \times r_k$ squares without remainder.

With this partition, the sum of side lengths of all obtained squares equals

$$q_1 n + q_2 r_1 + q_3 r_2 + \cdots + q_{k+1} r_k$$

$$= (m - r_1) + (n - r_2) + (r_1 - r_3) + \cdots + (r_{k-2} - r_k) + r_{k-1}$$

$$= m + n - r_k = m + n - (m, n).$$

Thus $f(m, n) \leqslant m + n - (m, n)$.

On the other hand, by the mathematical induction, we prove that for any partition satisfying the condition, the sum $b_{m, n}$ of side lengths of all squares is greater than or equal to $m + n - (m, n)$.

Without loss of generality, assume that $m \geqslant n$. If $m = 1$, then $n = 1$, and in this case, there is only one square. Hence the sum of side lengths of all square equals $1 = m + n - (m, n)$. Obviously,

$$b_{1,1} \geqslant 1 = m + n - (m, n).$$

Suppose that when $m \leqslant k$, for any $1 \leqslant n \leqslant m$, $b_{m,n} \geqslant m + n - (m, n)$ is valid. When $m = k + 1$, if $n = k + 1$, obviously, we obtain $b_{m,n} \geqslant k + 1 = m + n - (m, n)$. If $1 \leqslant n \leqslant k$, suppose that the $m \times n$ rectangle is divided into p squares with side lengths a_1, a_2, ..., a_p ($a_1 \geqslant a_2 \geqslant \cdots \geqslant a_p$) respectively. Obviously, $a_1 \leqslant n$.

(1) If $a_1 < n$, assume that in the rectangle $ABCD$, $AB = CD = m$, $BC = AD = n$. If the straight line l between AD and BC is parallel line of AD, then l passes through at least two squares (or l coincides with one sides of these squares). Hence

$$a_1 + a_2 + \cdots + a_p \geqslant AB + CD = 2m \geqslant m + n \geqslant m + n - (m, n).$$

(2) If $a_1 = n$, we remove an $n \times n$ square from the $m \times n$ rectangle and remain an $(m - n) \times n$ rectangle which is divided into $p - 1$ squares with side lengths a_2, a_3, ..., a_p respectively. By the inductive hypothesis, we obtain

$$a_2 + a_3 + \cdots + a_p \geqslant (m - n) + n - (m - n, n) = m - (m, n).$$

So $a_1 + a_2 + \cdots + a_p \geqslant m + n - (m, n)$.

With (1) and (2), it implies that

$$b_{m,n} \geqslant a_1 + a_2 + \cdots + a_p \geqslant m + n - (m, n).$$

Summarizing what we have described above, we know

$$f(m, n) = \min b_{m,n} = m + n - (m, n).$$

Remark   If there are several especial rectangles whose side lengths are given positive integers and we perform some especial partitions, then we may guess the required minimum in general case. Then, the inequality $f(m, n) \leqslant m + n - (m, n)$ is also proved by the mathematical induction method.

Example 3   Let A be a set of 200 different positive integers, and any three different positive integers of A be the lengths of three sides of a non-obtuse triangle. In this case, we say that the triangle is determined by the set A, and the sign $S(A)$ denote the sum of perimeters of all different triangles which are determined by the set A. Find the minimum of $S(A)$. (Two congruent triangles are regarded as the same).

Solution   Let $A = \{a_1, a_2, \ldots, a_{200}\}, a_1 < a_2 < \cdots < a_{200}, a_i \in N_+ (1 \leqslant i \leqslant 200)$. We know $a_i, a_j, a_k (1 \leqslant i < j < k \leqslant 200)$ are the lengths of three sides of a non-obtuse triangle if and only if $a_i^2 + a_j^2 \geqslant a_k^2$. It is equivalent to $a_1^2 + a_2^2 \geqslant a_{200}^2$. Thus

$$a_1^2 \geqslant a_{200}^2 - a_2^2$$

$$\geqslant (a_2 + 198)^2 - a_2^2$$

$$= 396a_2 + 198^2$$

$$\geqslant 396(a_1 + 1) + 198^2,$$

i.e. $a_1^2 - 396a_1 - 39600 \geqslant 0$. Hence

$$a_1 \geqslant 198 + \sqrt{198^2 + 39600} \geqslant 478.7,$$

so $a_1 \geqslant 479$, and it implies that $a_i \geqslant 478 + i \ (i = 1, 2, \ldots, 200)$.

Let $A_0 = \{479, 480, 481, \ldots, 677, 678\}$, then when $A = \{a_1, a_2, \ldots, a_{200}\} \ (a_1 < a_2 < \cdots < a_{200})$ satisfies the condition of the problem, we need $a_i \geqslant 478 + i \ (i = 1, 2, \ldots, 200)$. Thus $S(A) \geqslant S(A_0)$. On the other hand, obviously, $A_0 = \{479, 480, 481, \ldots, 677, 678\}$ satisfies the condition of the problem. (Since for any three numbers $a_i, a_j, a_k (a_i < a_j < a_k) \in A_0$, we know $a_i^2 + a_j^2 \geqslant 479^2 + 480^2 > 678^2 \geqslant a_k^2$.) Hence the minimum of $S(A)$ equals

$$S(A_0) = \sum_{\substack{a, b, c \in A_0 \\ a \neq b, b \neq c, c \neq a}} (a + b + c).$$

For any $a \in A_0$, the number of ways to select b, c is $\binom{199}{2}$, hence a is counted $\binom{199}{2}$ times in $S(A_0)$. Since a is any number of A_0, every number of A_0 is counted $\binom{199}{2}$ times in $S(A_0)$. Therefore

$$S(A_0) = \binom{199}{2} \sum_{a \in A_0} a$$

$$= \frac{1}{2} \times 199 \times 198 \times \sum_{i=479}^{678} i$$

$$= 2\ 279\ 405\ 700.$$

So that the required minimum of $S(A)$ is $2\ 279\ 405\ 700$.

Example 4   Find the greatest positive integer n such that there are n different real numbers x_1, x_2, $\ldots$, x_n satisfying that: for any $1 \leqslant i < j \leqslant n$, the inequality

$$(1 + x_i x_j)^2 \leqslant 0.99(1 + x_i^2)(1 + x_j^2)$$

holds.

Analysis

$$(1 + x_i x_j)^2 \leqslant 0.99(1 + x_i^2)(1 + x_j^2)$$
$$\Leftrightarrow 100(1 + x_i x_j)^2 \leqslant 99(1 + x_i^2)(1 + x_j^2)$$
$$\Leftrightarrow 99[(1 + x_i^2)(1 + x_j^2) - (1 + x_i x_j)^2] \geqslant (1 + x_i x_j)^2$$
$$\Leftrightarrow 99(x_i - x_j)^2 \geqslant (1 + x_i x_j)^2$$
$$\Leftrightarrow \left| \frac{x_i - x_j}{1 + x_i x_j} \right| \geqslant \frac{1}{\sqrt{99}}.$$

Thus naturally, we associate the left side of the above inequality with the formula from trigonometry:

$$\tan(\alpha - \beta) = \frac{\tan \alpha - \tan \beta}{1 + \tan \alpha \tan \beta}.$$

Solution   For n different real numbers x_1, x_2, $\ldots$, x_n, assume that $x_i = \tan \theta_i \left(-\frac{\pi}{2} < \theta_i < \frac{\pi}{2} \quad i = 1, 2, \ldots, n \right)$, and without loss of generality, assume that $\theta_1 < \theta_2 < \cdots < \theta_n$. Thus when $\theta_n - \theta_1 > \frac{n-1}{n}\pi = \pi - \frac{\pi}{n}$, with $\theta_n - \theta_1 < \pi$, we obtain

$$\tan^2(\theta_n - \theta_1) < \tan^2 \frac{\pi}{n}.$$

When $\theta_n - \theta_1 \leqslant \dfrac{n-1}{n}\pi$, with

$$\theta_n - \theta_1 = (\theta_n - \theta_{n-1}) + (\theta_{n-2} - \theta_{n-2}) + \cdots + (\theta_2 - \theta_1),$$

we obtain that there exists an i $(1 \leqslant i \leqslant n-1)$ such that $0 < \theta_{i+1} - \theta_i \leqslant \frac{\pi}{n}$. It implies that $\tan^2(\theta_{i+1} - \theta_i) \leqslant \tan \frac{\pi}{n}$. From above conclusion we know that there always exist $1 \leqslant i < j \leqslant n$ such that

$$\tan^2(\theta_j - \theta_i) \leqslant \tan^2 \frac{\pi}{n},$$

i. e.

$$\left(\cos^2 \frac{\pi}{n}\right)\left(\frac{x_j - x_i}{1 + x_j x_i}\right)^2 \leqslant \sin^2 \frac{\pi}{n}$$

$$\Leftrightarrow \left(\sin^2 \frac{\pi}{n}\right)(1 + x_j x_i)^2 \geqslant \left(\cos^2 \frac{\pi}{n}\right)(x_j - x_i)^2.$$

Adding $\left(\cos^2 \dfrac{\pi}{n}\right)(1 + x_j x_i)^2$ in the two sides, we yield

$$(1 + x_j x_i)^2 \geqslant \left(\cos^2 \frac{\pi}{n}\right)[(x_j - x_i)^2 + (1 + x_j x_i)^2]$$

$$= \left(\cos^2 \frac{\pi}{n}\right)(1 + x_i^2)(1 + x_j^2),$$

and when $n \geqslant 32$,

$$\cos^2 \frac{\pi}{n} = 1 - \sin^2 \frac{\pi}{n} \geqslant 1 - \left(\frac{\pi}{n}\right)^2 > 1 - \left(\frac{1}{10}\right)^2 = 0.99.$$

Hence when $n \geqslant 32$, then for n different real numbers x_1, $x_2, \ldots, x_n$, there are two real numbers x_i, x_j satisfying that

$$(1 + x_i x_j)^2 > 0.99(1 + x_i^2)(1 + x_j^2).$$

So when $n \geqslant 32$, there are not n different real numbers x_1, $x_2, \ldots, x_n$ such that for any $1 \leqslant i < j \leqslant n$, the inequality

$$(1 + x_i x_j)^2 \leqslant 0.99(1 + x_i^2)(1 + x_j^2)$$

holds.

On the other hand, we take 31 real numbers $x_i = \tan(i\theta)$ ($i = 1$, $2, \ldots, 31$), where $\theta = \arctan \dfrac{1}{\sqrt{99}}$, then

$$\tan\theta = \frac{1}{\sqrt{99}} = \frac{\sqrt{99}}{99} < \frac{10}{99} < \frac{\pi}{31} \Rightarrow 0 < \theta < \frac{\pi}{31}.$$

It implies that when $1 \leqslant i < j \leqslant 31$, then

$$\theta \leqslant (j - i)\theta \leqslant 30\theta < \frac{30\pi}{31} = \pi - \frac{\pi}{31} < \pi - \theta.$$

Thus for any $1 \leqslant i < j \leqslant 31$, we obtain

$$\tan^2(j - i)\theta \geqslant \tan^2\theta$$

$$\Leftrightarrow \left(\frac{x_j - x_i}{1 + x_j x_i}\right)^2 \geqslant \frac{1}{99}$$

$$\Leftrightarrow (1 + x_i x_j)^2 \leqslant 0.99(1 + x_i^2)(1 + x_j^2).$$

So there exist 31 different real numbers $x_1, x_2, \ldots, x_{31}$ which satisfy the condition of the problem. Therefore the required maximum of n is 31.

Example 5   Let $M = \{1, 2, \ldots, 40\}$. Find the smallest positive integer n for which it is possible to partition M into n disjoint subsets such that whenever a, b and c (not necessarily distinct) are in the same subset, $a \neq b + c$. (Belarus Mathematical Olympiad in 2000)

Solution   If $n = 4$, then we divide the set M into 4 disjoint subsets:

$$A = \{5, 6, 7, 8, 9, 32, 33, 34, 35, 36\},$$

$$B = \{14, 15, 16, \ldots, 25, 26, 27\},$$

$$C = \{2, 3, 11, 12, 29, 30, 38, 39\},$$

$$D = \{1, 4, 10, 13, 28, 31, 37, 40\},$$

thus for any three numbers a, b, c (not necessarily distinct) which belong to the same subset, and $a \neq b + c$. Hence the required smallest positive integer $n \leqslant 4$. Next, Assume, for sake of contradiction, that it is possible to partition M into 3 subsets A, B and C with the desired property. Without loss of generality, assume that $|A| \geqslant |B| \geqslant |C|$, thus

$$3 |A| \geqslant |A| + |B| + |C| = 40,$$

and $|A| \geqslant 14$. Assume that $a_1 < a_2 < \cdots < a_{14}$ are the elements of A. Thus each difference $b_i = a_{14} - a_i$ $(1 \leqslant i \leqslant 13)$ is in $B \cup C$ but cannot be in A, otherwise $(a_i) + (a_{14} - a_i) = a_{14}$, and it is a contradiction. By the Pigeonhole Principle, among 13 numbers $b_1 > b_2 > \cdots > b_{13}$, there are at least $\left[\dfrac{13 - 1}{2} \right] + 1 = 7$ belonging to B or C.

Without loss of generality, assume that b_1, b_2, ..., $b_7 \in B$. Similarly, each of $b_1 - b_2$, $b_1 - b_3$, $b_2 - b_3$, cannot be in A or B. Therefore the differences $b_1 - b_2 = a_2 - a_1$, $b_1 - b_3 = a_3 - a_1$, and $b_2 - b_3 = a_3 - a_2$ are in C. However, setting $a = a_3 - a_1$, $b = a_2 - a_1$, $c = a_3 - a_2$, we obtain $a = b + c$ and a, b, $c \in C$, and it is a contradiction. Therefore, it is impossible to partition M into three disjoint subsets with the desired property. Summarizing what we have described above, the required smallest positive integer n equals 4.

Example 6 Let $S = \{1, 2, 3, 4, 5, 6, 7, 8, 9, 10\}$, A_1, A_2, ..., A_k be the subsets of S satisfying the following conditions: (1) $|A_i| \geqslant 5$ $(1 \leqslant i \leqslant k)$; (2) $|A_i \cap A_j| \leqslant 2$ $(1 \leqslant i < j \leqslant k)$. Find the largest value of k. (The Examination Problem for The National Team of China for 35th IMO)

Solution I We construct a $10 \times k$ table such that the number in the ith row and the jth column is

$$x_{ij} = \begin{cases} 1, & \text{if } i \in A_j, \\ 0, & \text{if } i \notin A_j, \end{cases} \quad (i = 1, 2, \ldots, 10; j = 1, 2, \ldots, k).$$

Let $r_i = \sum_{j=1}^{k} x_{ij}$, $l_j = \sum_{i=1}^{10} x_{ij}$, then r_i respects that i belongs to r_i sets of A_1, A_2, ..., A_k and l_j equals the number of elements of A_j,

i. e. $l_j = |A_j|$. With the given condition (1), we know

$$\sum_{i=1}^{10} r_i = \sum_{i=1}^{10} \sum_{j=1}^{k} x_{ij} = \sum_{j=1}^{k} \sum_{i=1}^{10} x_{ij} = \sum_{j=1}^{k} |A_j| \geqslant 5k. \qquad ①$$

If $i \in A_t \cap A_j$ $(t \neq j)$, then i and A_t, A_j form a triple $\{i; A_t, A_j\}$ and let S denote the number of these triples.

On one hand, since i belongs to r_i sets of $A_1, A_2, \ldots, A_k$, there are $\binom{r_1}{2}$ triples $\{i; A_t, A_j\}$ containing i (with the convention $\binom{0}{1} = \binom{1}{2} = 0$). Hence

$$S = \sum_{i=1}^{10} \binom{r_i}{2} = \frac{1}{2} \left(\sum_{i=1}^{10} r_i^2 - \sum_{i=1}^{10} r_i \right). \qquad ②$$

On the other hand, for any two sets A_t, A_j $(t \neq j)$, there are $|A_t \cap A_j|$ elements belonging to $A_t \cap A_j$, hence there are $|A_t \cap A_j|$ triples $\{i; A_t, A_j\}$ containing A_t, A_j, and it implies that $S = \sum_{1 \leqslant t < j \leqslant k} |A_t \cap A_j|$. With the given condition (2), we obtain

$$S = \sum_{1 \leqslant t < j \leqslant k} |A_t \cap A_j| \leqslant 2\binom{k}{2} = k(k-1). \qquad ③$$

With ②, ③ and Cauchy's inequality and ①, we know

$$k(k-1) \geqslant \sum_{1 \leqslant t < j \leqslant k} |A_t \cap A_j|$$

$$= \frac{1}{2} \left(\sum_{i=1}^{10} r_i^2 - \sum_{i=1}^{10} r_i^2 \right)$$

$$\geqslant \frac{1}{2} \left(\frac{1}{10} \left(\sum_{i=1}^{10} r_i \right)^2 - \sum_{i=1}^{10} r_i \right)$$

$$= \frac{1}{20} \left(\sum_{i=1}^{10} r_i \right) \left(\sum_{i=1}^{10} r_i - 10 \right)$$

$$\geqslant \frac{1}{20} \times 5k \times (5k - 10)$$

$$= \frac{5}{4} k(k-2),$$

so $k \leqslant 6$.

Next, the following 6 subsets of M satisfy the conditions (1) and (2) of the problem:

$A_1 = \{1, 2, 3, 4, 5\}$, $A_2 = \{1, 2, 6, 7, 8\}$, $A_3 = \{1, 3, 6, 9, 10\}$,

$A_4 = \{2, 4, 7, 9, 10\}$, $A_5 = \{3, 5, 7, 8, 10\}$, $A_6 = \{4, 5, 6, 8, 9\}$.

Summarizing what we have described above, we obtain the required greatest value of k equals 6.

Solution II   We may construct 6 subsets of M satisfying the conditions (1) and (2) of the problem as solution I. Thus $k \geqslant 6$.

Next, for sake of contradiction, assume that $k \geqslant 7$, then from condition (1), we obtain $\sum\limits_{i=1}^{k} \mid A_i \mid \geqslant 5k \geqslant 35$. Thus there exists an element of M which belongs to at least 4 subsets of $A_1, A_2, \ldots, A_k$. Without loss of generality, assume that r belongs to four subsets: A_1, A_2, A_3, A_4. Thus for any $1 \leqslant i < j < t \leqslant 4$,

$$\mid A_i \cap A_j \cap A_t \mid \geqslant \mid A_1 \cap A_2 \cap A_3 \cap A_4 \mid \geqslant 1.$$

By the inclusion – exclusion principle and the given conditions (1) and (2), we obtain

$$10 = \mid M \mid \geqslant \mid A_1 \cup A_2 \cup A_3 \cup A_4 \mid$$

$$= \sum_{i=1}^{4} \mid A_i \mid - \sum_{1 \leqslant i < j \leqslant 4} \mid A_i \cap A_j \mid + \sum_{1 \leqslant i < j < t \leqslant 4} \mid A_i \cap A_j \cap A_t \mid$$

$$- \mid A_1 \cap A_2 \cap A_3 \cap A_4 \mid$$

$$\geqslant 4 \times 5 - \binom{4}{2} \times 2 + \left(\binom{4}{3} - 1 \right) \times 1$$

$$= 20 - 12 + (4 - 1) = 11,$$

and it is a contradiction. Therefore, our original assumption is incorrect, and it is impossible that $k \leqslant 6$.

Summarizing what we have descried above, the required greatest value of k is 6.

Example 7   Let MO be a city in firmament which consists of 99 space stations, and there are many tube routes connecting any two space stations. Assume that among these routes, there are exactly 99 routes which are bidirectional main routes and the remaining routes are strictly unidirectional. If four space stations satisfy that there is a route such that along this route, we could arrive at any space station from the others, then the set of the four space stations is called the mutual group of four stations.

Please design a scheme such that the number of the mutual groups of four stations is the greatest and find the greatest number. (The 14th CMO)

Solution   We consider a more general case: there are n space stations and n routes which are bidirectional main routes where $n \geqslant 3$ is an odd and put $m = \frac{1}{2}(n-3)$. In this problem, $n = 99$ and $m = 48$.

(1) If in a group of four stations, there is a station, called A, such that the routes connecting this station and three other stations are unidirectional and the initial station of these routes is the station A, then this set of four stations is not a mutual group of four stations. We use the sign S to denote the set of such groups of four stations which are not the mutual through groups of four stations and use the sign T to denote the set of the remaining groups of four stations which are not the mutual groups of four stations. Thus the number of the mutual group of four stations equals

$$N_n = \binom{n}{4} - |S| - |T|.$$

Assume the space stations are labeled $1, 2, \ldots, n$ and the number of unidirectional routes with initial station i is x_i, then $|S| = \sum_{i=1}^{n} \binom{x_i}{3}$ and

$$x_1 + x_2 + \cdots + x_n = \binom{n}{2} - n = \frac{1}{2}n(n-3) = mn.$$

N_n attains the greatest value when $|S|$ and $|T|$ take the smallest values. Firstly, we prove that $|S| \geqslant n\binom{m}{3}$. In fact, if $|S|$ attains the smallest value, then for any $1 \leqslant i < j \leqslant n$, $| x_i - x_j | \leqslant 1$. If there exist $1 \leqslant i < j \leqslant n$ such that $x_j - x_i \geqslant 2$, then setting $x_i' = x_i + 1$, and $x_j' = x_j - 1$, $x_k' = x_k$ $(k \neq i, j)$, we obtain $\sum\limits_{i=1}^{n} x_i' = \sum\limits_{i=1}^{n} x_i$. Denote $| S' | = \sum\limits_{i=1}^{n} \binom{x_i'}{3}$, then

$$
\begin{aligned}
| S | - | S' | &= \binom{x_i}{3} + \binom{x_j}{3} - \binom{x_i'}{3} - \binom{x_j'}{3} \\
&= \binom{x_i}{3} + \binom{x_j}{3} - \binom{x_i + 1}{3} - \binom{x_j - 1}{3} \\
&= \binom{x_j - 1}{3} - \binom{x_i}{3} > 0
\end{aligned}
$$

(since $(x_j - 1) - x_i \geqslant 1$), and it contradicts that $| S |$ attains the smallest value. Since $x_1 + x_2 + \cdots + x_n = mn$, when $x_1 = x_2 = \cdots = x_n = m$, $| S |$ attains the smallest value, and it implies that $| S | \geqslant n \binom{m}{3}$, so the equality holds if and only if $x_1 = x_2 = \cdots = x_n = m$. Therefore

$$
\begin{aligned}
N_n &= \binom{n}{4} - | S | - | T | \\
&\leqslant \binom{n}{4} - n \binom{m}{3} - 0 \\
&= \frac{1}{18} n(n - 3)(n^2 + 6n - 31). \qquad \qquad ①
\end{aligned}
$$

(2) Next, the following scheme indicates that the equality in ① holds.

Firstly, we arrange the space stations clockwise with serial numbers $1, 2, \ldots, n$ at n points $A_1, A_2, \ldots, A_n$ in a circle. Assume that the routes between two adjacent stations in this circle are the bidirectional main routes, i. e. the bidirectional main routes are the following n routes:

$$
A_1 A_2, A_2 A_3, \ldots, A_{n-1} A_n, A_n A_1.
$$

For any i, $j \in \{1, 2, \ldots, n\}$, $i \neq j$, if the arc $\overset{\frown}{A_iA_j}$ from A_i to A_j in clockwise contains odd points, we definite that the route between A_i and A_j is strictly unidirectional from A_i to $A_j : A_i \rightarrow A_j$. Since n is an odd, then just one of the arcs from A_i to A_j and from A_j to A_i in clockwise contains odd points. Hence the above definition has not the contradiction.

According to the above definition, the number of the strictly unidirectional routes from A_i to other points equals $m = \dfrac{1}{2}(n - 3)$, hence $|S| = n\dbinom{m}{3}$. We will prove that $|T| = 0$ in this scheme.

If there the two space stations in a group of four stations such that the route between them is a bidirectional main route, then we know that this group of four stations is a mutual group of four stations. Hence if the group of four stations A, B, C, D is not a mutual group of four stations, then the routes between any two stations of A, B, C, D are strictly unidirectional. Assume the numbers of space stations between A and B, B and C, C and D, and D and A are a, b, c, d respectively, then $a + b + c + d = n - 4$ is an odd. It implies that the number of odds in a, b, c, d equals 1 or 3.

(I) If a is odd and b, c, d are even then $A \rightarrow B \rightarrow D \rightarrow C \rightarrow A$, i.e. A, B, C, D construct a mutual group of four stations. (As in figure 15. 1)

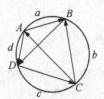
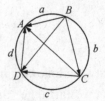

| Figure 15. 1 | Figure 15. 2 |

(II) If a is even and b, c, d are odd, and the routes from B to A, C, or D are strictly unidirectional routes with the initial station B and such group of four stations is not a mutual group of four stations which belonging set S. (As in figure 15. 2)

From the above discussion, we obtain that $|T| = 0$ in this scheme, hence

$$|S| = \binom{n}{4} - n\binom{m}{3}.$$

Summarizing what we have described above, we obtain that the greatest value of number of mutual groups of four stations equals

$$\binom{n}{4} - n\binom{m}{3} - 0 = \frac{1}{18}n(n-3)(n^2 + 6n - 31).$$

Especially, when $n = 99$, we conclude that the greatest number of mutual groups of four stations in this problem equals

$$\binom{99}{4} - 99\binom{48}{3} = 2\,052\,072.$$

Example 8 Let the set $U = \{1, 2, \ldots, n\}$, where $n \geqslant 3$. A subset S of U is said to be split by an arrangement of elements of U if an element not in S occurs in the arrangement somewhere between two elements of S. For example, 13452 splits $\{1, 2, 3\}$ but not $\{3, 4, 5\}$. Find the greatest positive integer m such that for any m subsets, each containing at least 2 and at most $n - 1$ elements of U, there is an arrangement of U which splits all of the m subsets (The Problem Prepared for 39th IMO in 1998).

Solution For $n = 3$, there is not an arrangement of $\{1, 2, 3\}$ which splits two subset $\{1, 2\}$ and $\{1, 3\}$ simultaneously, but for any a 2-element subset $A = \{i, j\}$ of $\{1, 2, 3\}$, there is an arrangement $\{i, k, j\}$ of $\{1, 2, 3\}$ which splits A. Hence when $n = 3$, the greatest value of m equals 1. Similarly, when $n = 4$, we may obtain that the greatest value of m equals 2. In general case, we prove that the required greatest value of m equals $n - 2$.

Firstly, by mathematical induction method, we prove that for any $n - 2$ subsets in $U = \{1, 2, \ldots, n\}$, there exists an arrangement of U which splits these $n - 2$ subsets.

In fact, for the basic case $n = 3$, we have proved the conclusion is true. Assume for $n \geqslant 3$, the conclusion is true. Assume $U = \{1, 2, \ldots, n, n + 1\}$ and $\Re$ is a family of $n - 1$ subsets of U, each containing at least 2 and at most n elements of U. Suppose that $\Re$ contains k 2-element subsets and l n-element subsets. Then $k + l \leqslant n - 1$. Since at most k elements of U can appear two or more times in the k 2-element subsets. Hence, the number of elements which appear at most once among the k 2-element subsets is at least

$$(n + 1) - k \geqslant (n + 1) - (n - 1 - l) = l + 2.$$

If $l = 0$, then there is not a n-element subset in $\Re$. If $l \geqslant 1$, then there are just l elements which are not contained in some of the l n-element subsets. (Since for each n-element subset in U, there is only one element of U which does not belong to it.) Hence among the $l + 2$ elements, there is an element belong to all the n-element subsets of U. Thus we have proved that there is an element a in each of n-element subsets of U, but at most one of the 2-element subsets of U.

Without loss of generality, assume that $a = n + 1$, otherwise we rearrange the elements of U. When the element $n + 1$ is removed, all n-element subset in U become $(n - 1)$-element subsets of $\{1, 2, \ldots, n\}$ (or there is not an n-element subset of U), and there are at most one of the 2-element subset becomes a singleton.

If we have such a subset $\{i\}$, then the induction hypothesis guarantees the existence of a permutation π of $\{1, 2, \ldots, n\}$ that splits all the other $n - 2$ subsets (with $n + 1$ taken out). By adding $n + 1$ to π anywhere away from i, we have a permutation of $\{1, 2, \ldots, n + 1\}$ which splits all $(n - 1)$-element subsets in U. (Note that all of the other subsets that contain $n + 1$ which were already split before adding $n + 1$ remain split.)

If we do not have such a singleton, choose any subset S among the $(n - 1)$-element subsets. By the induction hypothesis, we have a permutation π of $\{1, 2, \ldots, n\}$ that splits all the other $n - 2$ subsets. If $n + 1 \notin S$, adding $n + 1$ to π between two elements of S, we have a permutation of $\{1, 2, \ldots, n + 1\}$ which splits all $(n - 1)$-element subsets in U. Otherwise, if $n + 1 \in S$, if π does not already split S, we may add $n + 1$ on either the left or right end of π to split S. This completes our induction.

On the other hand, for any $n \geqslant 3$, there is not a permutation of $\{1, 2, \ldots, n\}$ which split all of following $n - 1$ subsets: $\{1, 2\}$, $\{1, 3\}$, $\{1, 4\}$, $\ldots$, and $\{1, n\}$. In fact, if the permutation π splits the subsets: $\{1, 2\}$, $\{1, 3\}$, $\{1, 4\}$, $\ldots$, and $\{1, n - 1\}$, then π just has the two following forms: $\pi_1 = \{1na_2a_3\cdots a_{n-1}\}$ or $\pi_2 = \{a_2a_3\cdots a_{n-1}n1\}$, where $\{a_2a_3\cdots a_{n-1}\}$ is any permutation of 2, 3, $\ldots$, n, but π_1 and π_2 cannot split the subset $\{1, n\}$.

Summarizing what we have described above, the required greatest value of m equals $n - 2$.

Exercise 15

1 Assume that for a finite set A, there exists a function $f: \mathbf{N}_+ \rightarrow A$ which has following property: If i, $j \in \mathbf{N}_+$ and $|i - j|$ is a prime, then $f(i) \neq f(j)$. What is the smallest number of elements that A can have?

2 M is a finite set of real numbers such that given three distinct elements from M, we can choose two of them whose sum also belongs to M. What is the largest number of elements that M can have? (26th Russian Mathematical Olympiad in 2000)

3 Let $M = \{1, 2, 3, \ldots, 19, 20\}$, and A_1, A_2, $\ldots$, A_n be different non-empty subsets of M such that when $i \neq j$, $|A_i \cap A_j| \leqslant 2$. Find the largest value of n.

4 Given positive integer $n \geqslant 3$, find the smallest positive integer m satisfying that if we partition the set $S = \{1, 2, \ldots, m\}$ into any two disjoint subsets, then there are n positive integers x_1, x_2, $\ldots$, x_n (not necessarily distinct) belonging to the same subset such that $x_1 + x_2 + \cdots + x_{n-1} = x_n$.

5 In a mathematical competition, there are 8 players and n two -

choice (yes or no) questions. Assume for any two questions (A, B) (where (A, B) is an order pair), there are exactly two persons whose answers are (yes, yes), two persons (yes, no), two persons (no, yes) and two persons (no, no). Find the largest value of n. Justify your answer.

6 Let set $S = \{1, 2, 3, \ldots, 2005\}$, and $A = \{a_1, a_2, \ldots, a_k\}$ be a subset of S, and the sums of any two numbers in A are not a multiple of 125. Find the largest value of k.

7 Let $P = \{P_1, P_2, \ldots, P_{1994}\}$ be a set of the planar points, no three points of which are collinear. The set P is partitioned into 83 subsets such that there are at least three points in each subset. Assume that any two points in the same subset are connected by a line segment and any two points in distinct subset are not connected by a line segment. Thus we get a figure G, and the distinct partitions correspond to distinct figures. Let the sign $m(G)$ denote the number of triangles whose three vertices are the points in P. (1) Find the smallest value m_0 of $m(G)$; (2) Assume that figure G^* is a figure such that the equality $m(G^*) = m_0$ holds. If the line segments in G^* are colored with four distinct colors and every line segment is colored with a color. Prove that there is a coloring scheme such that when G^* is colored, and there is not a triangle whose three side have the same color and its vertices is in P. (China Mathematical Competition in 1994)

8 There are 2000 circular pieces of paper with the same radii without overlapping on a table, and some circles are circumscribed. Find the smallest number of colors such that each circular piece of paper is colored with a color, and any two circles which are circumscribed are colored with distinct colors. (The Selective Examination for National Team of China for the 30th IMO)

9 For integer $n \geqslant 4$, find the minimal integer $f(n)$, such that for any positive integer m, in any subset with $f(n)$ elements of the set $\{m, m+1, m+2, \ldots, m+n-1\}$ there are at least 3 mutually prime elements. (China Mathematical Competition in 2004)

Solutions to Exercises

Exercise 1

1 With the given condition, we know that there is exactly one student who gets 2 prizes and each of the remaining $n-1$ students who gets exactly 1 prize, then the number of distinct ways to award prizes is $\binom{n+1}{2}P_n^n$. Answer: D.

2 The number of distinct of selecting ways is $\binom{5}{1}\binom{4}{3}+\binom{5}{2}\binom{4}{2}+\binom{5}{3}\binom{4}{1}=120$. Answer: C.

3 Firstly, the unit digit of 5-digit numbers which are not the multiple of 5 is one of 1, 2, 3, 4. Thus the number of such 5-digit numbers is $4\times4!$, among these numbers, the number of 5-digit numbers not exceeding 20000 whose leading digit is 1 and unit digit is one of 2, 3, 4 is $3\times3!$. Hence the number of 5-digit numbers satisfying the conditions of the problem is $4\times4!-3\times3!=78$. Answer: B.

4 The number of ways of selecting 2 numbers A and B from 6 digits is P_2^6. But in the above computation, the straight lines corresponding to the following cases are the same straight line: (1) $A=0$, $B=1,2,3,6,7$; (2) $B=0, A=1,2,3,6,7$; (3) $A=1, B=2$ and $A=3, B=6$; (4) $A=1, B=3$ and $A=2, B=6$; (5) $A=2, B=1$ and $A=6, B=3$; (6) $A=3, B=1$ and $A=6, B=2$. Hence the number of distinct straight lines is $P_2^6-2(P_1^5-1)-4=18$.

5 The number of ways of selecting 2 numbers a and x from 7 digits 1, 2, 3, 4, 5, 7, 9 is P_2^7. But the base a can not equal 1 and the countings are repeated in the following cases: (1) $x = 1$ and $a = 2, 3,$ 4, 5, 7, 9; (2) $\log_2 4 = \log_3 9$; (3) $\log_4 2 = \log_9 3$; (4) $\log_3 2 = \log_9 4$; (5) $\log_2 3 = \log_4 9$. Hence the number of distinct values of $\log_a x$ is

$$P_2^7 - \binom{6}{1} - \left(\binom{6}{1} - 1\right) - 4 = 27.$$

6   Assume that there are n players and the number of games among the 3 players is r. Thus $50 = \binom{n-3}{2} + (3 \times 2 - r)$, this is $(n - 3)(n - 4) = 88 + 2r$. Thus $r = 1$, $n = 13$. Answer: **B**.

7   suppose that the inclination is θ, then $\tan \theta = -\dfrac{a}{b} > 0$. Without loss of generality, assume that $a > 0$, $b < 0$. (1) When $c = 0$, we choose a in $\binom{3}{1}$ ways and b in $\binom{3}{1}$ ways respectively. But we ought to delete two repeated cases (Since three equation $x - y = 0$, $2x - 2y = 0$, $3x - 3y = 0$ correspond to a straight line). Hence in this case, the number of distinct straight lines is $\binom{3}{1}\binom{3}{1} - 2 = 7$; (2) When $c \neq 0$, we choose a in $\binom{3}{1}$ ways, b in $\binom{3}{1}$ ways and c in $\binom{4}{1}$ ways respectively. Hence in this case, the number of distinct straight lines is $\binom{3}{1}\binom{3}{1}\binom{4}{1} = 36$. Therefore the number of straight lines satisfying the conditions is $7 + 36 = 43$.

8   As figure, A and B are colored in $P_2^6 = 30$ ways. If C and B are colored with the same color, then C and D are colored in $\binom{5}{1}$ ways; If C and B are colored with distinct colors, then C and D are colored in $\binom{4}{1}\binom{4}{1}$ ways. Hence C and D are colored in $\binom{5}{1} + \binom{4}{1}\binom{4}{1} = 21$ ways. Similarly, E and F are colored

A	C	E
B	D	F

in 21 ways. By the multiplication principle we obtain that the number of distinct coloring ways satisfying the conditions is $30 \times (21)^2 = 13\,230$.

9 Firstly, we find the number of distinct processes of the game A wins. Assume that i^{th} player of A wins x_i times $(i = 1, 2, \ldots, 7)$, then $x_1 + x_2 + \cdots + x_7 = 7$. Thus the number of distinct processes of game A wins equals the number of nonnegative integer solutions of the indefinite equation $x_1 + x_2 + \cdots + x_7 = 7$, and it is $\binom{7+7-1}{7-1} = \binom{13}{6}$. Similarly, the number of distinct processes of the game B wins also equals $\binom{13}{6}$. So the number of distinct processes of the game equals

$$2\binom{13}{6} = 3\,432.$$

10 Since the marksman break eight targets in $8!$ ways and the orders of the targets in every column are determined, the required number of ways is $\dfrac{8!}{2! \cdot 2! \cdot 3!} = 560$.

11 Obviously, we need at least three colors. (1) If 5 colors are all used, then we take one color from 5 colors to color the above vertex in $\binom{5}{1}$ ways. Afterward, we use the remaining 4 colors to color 4 vertices in base facet in $(4-1)! = 3!$ ways. In this case, there are, in all, $\binom{5}{1} \cdot 3! = 30$ ways; (2) If 4 colors are used, then we choose 4 colors from 5 colors in $\binom{5}{4}$ ways and use one of the 4 colors to color the above vertex in $\binom{4}{1}$ ways. Afterward, we use the remaining 3 colors to color 4 vertices in base facet. In this case, one pair of the opposite vertices in base facet is colored with the same color. We choose a color to color one pair of the opposite vertices in base facet in

$\binom{3}{1}$ ways, and use the remaining 2 colors to color the remaining two vertices in base facet in $1! = 1$ way. In this case, there are, in all, $\binom{5}{4}\binom{4}{1}\binom{3}{1} \cdot 1 = 60$ ways; (3) If 3 colors are used, we take 3 colors from 5 colors in $\binom{5}{3}$ ways and use one of the 3 colors to color the above vertex in $\binom{3}{1}$ ways. Afterward, we use the remaining 2 colors to color 4 vertices in base facet. In this case, each pair of the opposite vertices in base facet is colored with the same color, and the number of the coloring ways is $1! = 1$. In this case, there are, in all, $\binom{5}{3}\binom{3}{1} \cdot 1! = 30$ ways. Summing up the above, we obtain that the number of ways satisfying conditions is $30 + 60 + 30 = 120$.

12 We might as well suppose $b_1 < b_2 < \cdots < b_{50}$, and divide elements $a_1, a_2, \ldots, a_{100}$ in A into 50 nonempty groups according to their order. Define a mapping $f : A \rightarrow B$, so that the images of all the elements in the ith group are $b_i (i = 1, 2, \ldots, 50)$ under the mapping. Obviously, f satisfies the requirements given in the problem. Furthermore, there is a one-to-one correspondence between all groups so divided and the mappings satisfying the condition. So the number of mappings f satisfying the requirements is equal to the number of ways of dividing A into 50 groups according to the order of the subscripts. Let x_i denote the number of elements of the ith group, then the number of ways of dividing A is equal to the number of positive integer solutions of the indefinite equation $x_1 + x_2 + \cdots + x_{50} = 100$, and it is $\binom{100-1}{50-1} = \binom{99}{49}$. Then there are, in all, $\binom{99}{49}$ such mappings. Answer: D.

13 Since the number of non-negative integer solutions of equation $x_1 + x_2 + \cdots + x_k = m$ is $\binom{m+k-1}{m}$, when $x_1 \geqslant 1$ and $x_i \geqslant$

0 $(i \geqslant 2)$, the number of integer solutions, is $\binom{m+k-2}{m-1}$, the

number of lucky numbers with k digits is $P(k) = \binom{k+5}{6}$. We know

2005 is the minimum lucky number of the type $\overline{2abc}$ and $P(1) = \binom{6}{6} =$

1, $P(2) = \binom{7}{6} = 7$, and $P(3) = \binom{8}{6} = 28$. Note that the number of

four digits lucky numbers of the type $\overline{1abc}$ is the number of

nonnegative integer solutions of $a+b+c = 6$, i.e. $\binom{6+3-1}{6} = 28$.

Thus $1+7+28+28+1 = 65$ and 2005 is the 65th lucky number, i.e. a_{65}

$= 2005$, so $n = 65$, $5n = 325$. Furthermore, $P(4) = \binom{9}{6} = 84$, $P(5) =$

$\binom{10}{6} = 210$, and $\sum_{k=1}^{5} P(k) = 330$. Therefore the last six lucky numbers

with 5 digits, from the largest to the smallest, are $a_{330} = 70000$, $a_{329} =$
61000, $a_{328} = 60100$, $a_{327} = 60010$, $a_{326} = 60001$, $a_{325} = 52000$. So the
325th lucky number is $a_{325} = a_{5n} = 52000$.

14 Let S denote the set of all permutations of the n pair married
couples in a line and A_i denote the set of all permutations in which the
i^{th} pair married couple is adjacent. By the inclusion-exclusion
principle, the number of all permutations satisfying the condition of
the problem equals

$$\left| \bigcap_{i=1}^{n} \overline{A_i} \right| = |S| - \sum_{i=1}^{n} |A_i| + \sum_{1 \leqslant i < j \leqslant n} |A_i \cap A_j| - \cdots + (-1)^n \left| \bigcap_{i=1}^{n} A_i \right|$$

$$= (2n)! - \binom{n}{1} \cdot 2 \cdot (2n-1)! + \binom{n}{2} \cdot 2^2 \cdot (2n-2)! - \cdots$$

$$+ (-1)^k \binom{n}{k} \cdot 2^k \cdot (2n-k)! + \cdots + \binom{n}{n} \cdot 2^n \cdot n!.$$

15 Let $a_{1000} = n$, $S = \{1, 2, \ldots, n\}$, and $A_i = \{k \mid k \in S$, and
k is a multiple of $i\}$ $(i = 3, 5, 7)$. Since $105 = 3 \times 5 \times 7$, then $(k,$

$105) \Leftrightarrow k$ is not multiple of 3, 5 and 7. By the inclusion-exclusion principle and the given conditions, we obtain that

$$1000 = | \overline{A_3} \cap \overline{A_5} \cap \overline{A_7} |$$
$$= | S | - | A_3 | - | A_5 | - | A_7 | + | A_3 \cap A_5 | + | A_3 \cap A_7 |$$
$$+ | A_5 \cap A_7 | - | A_3 \cap A_5 \cap A_7 |$$

$$= n - \left[\frac{n}{3} \right] - \left[\frac{n}{5} \right] - \left[\frac{n}{7} \right] + \left[\frac{n}{3 \times 5} \right] + \left[\frac{n}{3 \times 7} \right]$$
$$+ \left[\frac{n}{5 \times 7} \right] - \left[\frac{n}{3 \times 5 \times 7} \right]. \qquad \qquad \text{①}$$

With $\alpha - 1 < [\alpha] \leqslant \alpha$, and ①, we obtain that

$$1000 > n - \left(\frac{n}{3} + \frac{n}{5} + \frac{n}{7} \right) + \left(\frac{n}{3 \times 5} - 1 + \frac{n}{3 \times 7} - 1 + \frac{n}{5 \times 7} - 1 \right)$$
$$- \left(\frac{n}{3 \times 5 \times 7} \right)$$

and

$$1000 < n - \left(\frac{n}{3} - 1 + \frac{n}{5} - 1 + \frac{n}{7} - 1 \right) + \left(\frac{n}{3 \times 5} + \frac{n}{3 \times 7} + \frac{n}{5 \times 7} \right)$$
$$- \left(\frac{n}{3 \times 5 \times 7} - 1 \right).$$

It means $2178 \frac{3}{4} < n < 2194 \frac{1}{16}$. Since n and 105 are coprime, so n is one of 2 179, 2 182, 2 183, 2 186, 2 188, 2 189, 2 192, and 2 194. Among these numbers, there is only $n = 2\ 186$ which satisfies the equation ①, so $a_{1000} = 2\ 186$.

16 Let S denote the set of the n-digit numbers consisting of the digits 1, 2 and 3. $A_i = \{m \mid m \in S$, and each digit of m is not $i\}$ $(i = 1, 2, 3)$. Thus the number of required n-digit numbers equals

$$| \overline{A_1} \cap \overline{A_2} \cap \overline{A_3} | = | S | - | A_1 | - | A_2 | - | A_3 | + | A_1 \cap A_2 |$$
$$+ | A_1 \cap A_3 | + | A_2 \cap A_3 | - | A_1 \cap A_2 \cap A_3 |$$
$$= 3^n - (2^n + 2^n + 2^n) + (1^n + 1^n + 1^n) - 0$$
$$= 3^n - 3 \cdot 2^n + 3.$$

Exercise 2

1 Let 17 points $A_i \, (i = 1, 2, \ldots, 17)$ in a circle represent 17 scientists. If a couple of scientists makes reference to the first (second or third) topic, then the line segment connecting two corresponding points is colored red (blue or green). By Pigeonhole Principle, of 16 line segments meeting A_i at least $\left[\dfrac{16-1}{3}\right] + 1 = 6$ is colored with the same color. Without loss of generality, suppose that the line segments $A_1 A_2$, $A_1 A_3$, $\ldots$, $A_1 A_7$ are colored red. If among six points A_2, A_3, $\ldots$, A_7, there are two points A_i, $A_j \, (2 \leqslant i < j \leqslant 7)$ such that the line segment connecting the two points is colored red, then there is a red triangle $A_1 A_i A_j$. Otherwise, the line segments connecting each pair of six points A_2, A_3, $\ldots$, A_7 are all colored with blue or green, by the Ramsey's theorem, there exists a monochromatic triangle, i.e. there exists a group of three scientists whose letters are about the same topic.

2 (1) Numbers 1, 3, 7, and 9 satisfy the conditions of the problem. (2) Consider the remainders of any 5 positive integers with modulus 3. If three remainders 0, 1, and 2 all appear, then the sum of the three numbers is divisible by 3, so it is not a prime. If at more two of the three remainders 0, 1, and 2 occur, then by Pigeonhole Principle, at least $\left[\dfrac{5-1}{2}\right] + 1 = 3$ numbers whose remainders with modulus 3 are equal, hence the sum of the three numbers is divisible by 3, and it is not a prime. Summing up the above, there are not five numbers such that the sum of any three of these numbers is a prime.

3 As figure, the triangle ABC is divided into two triangles: $\triangle AEF$ and $\triangle CDE$ and a parallelogram $BDEF$. By Pigeonhole Principle, there exist at least $\left[\dfrac{7-1}{3}\right] + 1 = 3$ points in same figure. So the area of triangle whose vertices are this three points is less than or equal to $\dfrac{1}{4}$.

4 Assume that $M = \{3k + 1 \mid k = 0, 1, 2, \ldots, 669\}$, then $M \subset S$, $M \neq S$ and $|M| = 670$. Since the difference of any two numbers in M is a multiple of 3, and the sums of any two numbers in M is not a multiple of 3, the sum of any two numbers in M is not divisible by their difference. So the required largest value is not less than 670. Next, let S be partitioned into the following 670 subsets: $\{1, 2, 3\}$, $\{4, 5, 6\}$, $\{7, 8, 9\}$, $\ldots$, $\{2005, 2006, 2007\}$, $\{2008, 2009\}$. We take any 671 numbers from M, by Pigeonhole Principle, at least $\left[\dfrac{671 - 1}{670}\right] + 1 = 2$ belongs to the same subset and their difference does not exceed 2. Assume that the two numbers is a and b and $a > b$. If $a - b = 1$, then $a + b$ is divisible by $a - b$. If $a - b = 2$, then a and b are odds or evens. Hence $a + b$ is an even, and $a + b$ is also divisible by $a - b$. Therefore the required largest positive integer equals 670.

5 We first prove that given any set of 13 consecutive positive integers from S, at more 6 of these 13 integers can be elements of M. We prove this fact for the set $T = \{1, 2, 3, \ldots, 13\}$, but the same proof works for any set of 13 consecutive positive integers. Consider the following 13 subsets of T: $\{1, 6\}$, $\{2, 7\}$, $\{3, 8\}$, $\{4, 9\}$, $\{5, 10\}$, $\{6, 11\}$, $\{7, 12\}$, $\{8, 13\}$, $\{1, 9\}$, $\{2, 10\}$, $\{3, 11\}$, $\{4, 12\}$, and $\{5, 13\}$, then each number of T belongs to just two subsets. For any 7 numbers of T which belong to the above 14 subsets, and by Pigeonhole Principle, at least two numbers belong to the same subset whose difference equals 5 or 8. Hence any 7 numbers of T are not all to belong to M. On the other hand, it is easy to find a 6-element subset that satisfies the key property of M (i.e., no two numbers differ by 5 or 8). One such set is $T' = \{1, 2, 4, 5, 8, 11\}$. We also find that T' has the remarkable property of allowing for a periodic continuation. This is, if $\mathbf{Z}$ denotes the set of integers, then $M' = \{13n + k \mid n \in \mathbf{Z}, k = 1, 2, 4, 5, 8, 11\}$ also has the property that no two elements in the set differ by 5 or 8. Moreover, since $2000 = 13 \times 154 - 2$, so at most $154 \times 6 = 924$ numbers of M satisfy that no two number differ by 5 or 8. Thus set $M = M' \cap S$ has the property that no two elements in the set differ by 5 or 8 and $|M| = |M' \cap S| = 924$. Summing up the above, we get obtain that the largest number of elements of M is 924.

6 For sake of contradiction, assume that two persons A and B are not brothers, then each of the remaining 5 persons and at most one of $\{A, B\}$ are brothers. (Otherwise, three persons A, B, and C are brothers, and it is a contradiction.) By Pigeonhole Principle, one of $\{A, B\}$ has at most $\left[\dfrac{5}{2}\right] = 2$ brothers among the remaining 5 persons. Thus this person has at most 2 brothers among the remaining 6 persons, and it contradicts the given condition. This completes the proof.

7 Assume that the number of purchasers of some book is largest and it is k. If $k \leqslant 4$, since $4n \neq 30$ ($n \in \mathbf{N}_+$), then it is impossible that the numbers of purchasers of each book are all 4. Assume that some person P buys 3 books: a, b and c, and the sets of purchasers of a, b and c are M_a, M_b and M_c respectively, and $|M_a| \leqslant 3$, $|M_b| \leqslant 4$, $|M_c| \leqslant 4$.

Since $P \in M_a \cap M_b \cap M_c$, $|M_a \cap M_b| \geqslant 1$, $|M_a \cap M_c| \geqslant 1$, $|M_b \cap M_c| \geqslant |M_a \cap M_b \cap M_c| \geqslant 1$ and from the given condition (2), we know $|M_a \cup M_b \cup M_c| = 10$. By the inclusion-exclusion principle, we obtain that $10 = |M_a \cup M_b \cup M_c| = |M_a| + |M_b| + |M_c| - |M_a \cap M_b| - |M_a \cap M_c| - (|M_b \cap M_c| - |M_a \cap M_b \cap M_c|) \leqslant 3 + 4 + 4 - 1 - 1 - 0 = 9 - 1 - 1 - 0 = 9$, and it is a contradiction, thus $k \geqslant 5$. On the other hand, we denote books by the numbers: 1, 2, 3, ... and the books which are brought by the 10 persons are: $\{1, 2, 3\}$, $\{1, 2, 3\}$, $\{1, 4, 5\}$, $\{1, 6, 7\}$, $\{2, 4, 6\}$, $\{2, 4, 6\}$, $\{2, 5, 7\}$, $\{3, 4, 7\}$, $\{3, 5, 6\}$, and $\{3, 5, 6\}$ respectively, then the given conditions in the problem are satisfied and $k = 5$ holds. Therefore, the required smallest value is 5.

8 (1) Firstly, we prove that for any real number a, there exist two points such that the distance between the two points equals $2a$ and this two points are colored with the same color. In fact, we choose a red point A and draw a circle with the center A and radius $2a$. If there is a red point on this circle, then the conclusion holds. Otherwise, the points on this circle are all blue. Thus on this circle, two end points of one side of inscribed regular hexagon are all blue and the distance between the two points equals $2a$. Thus the conclusion also holds. Next, let $AB = 2a$ and A, B be red points. We draw a circle with diameter AB and denote 6 vertices of inscribed regular hexagon of this circle by A, C, D, B, E and F respectively. If one of C, D, E, F, say C, is a red point, then three vertices of $\mathrm{Rt}\triangle ABC$ are all red and $BC = a$, $CA = \sqrt{3}a$, and $AB = 2a$, so the conclusion is valid. Otherwise, three vertices of $\mathrm{Rt}\triangle CDF$ are all blue and $CD = a$, $DF = \sqrt{3}a$, and $EC = 2a$, so the conclusion is also valid.

(2) **Proof** I　Taking $a = 1$, and $a = 1995$ respectively in (1), we obtain two right-angle triangles satisfying the given conditions in the problem. This completes the proof.

Proof II　We draw two circles with the same center O, and choose any 9 points on the little circle. By Pigeonhole Principle, at least $\left[\dfrac{9-1}{2}\right]+1 = 5$ points of the 9 points are colored with the same color. Let 5 points A_1, A_2, A_3, A_4, A_5 on the little circle be colored red. We draw ray $OA_i (i = 1, 2, 3, 4, 5)$ and denote the intersecting points of OA_i and the greater circle by $B_i (i = 1, 2, 3, 4, 5)$. By Pigeonhole Principle, at least $\left[\dfrac{5-1}{2}\right]+1 = 3$ points B_{i_1}, B_{i_2}, and B_{i_3} $(1 \leqslant i_1 < i_2 < i_3 \leqslant 5)$ are colored with the same colors. Thus two triangles $\triangle A_{i_1} A_{i_2} A_{i_3}$ and $\triangle B_{i_1} B_{i_2} B_{i_3}$ satisfy the given conditions in the problem. This completes the proof.

9　Suppose the convex has F facets, E edges and V vertices, then $V = 6$, and $E = 12$. Substituting them to Euler's formula, we obtain that $F = E + 2 - V = 12 + 2 - 6 = 8$. Assume that there are x_i edges on the ith facet $(i = 1, 2, \ldots, 8)$, thus $x_i \geqslant 3$ and $\displaystyle\sum_{l=1}^{8} x_i = 2E = 24$, i.e.

$\frac{1}{8} \sum_{i=1}^{8} x_i = 3$. If there is some $x_i > 3$, then $\frac{1}{8} \sum_{i=1}^{8} x_i > 3$, and it is a contradiction. So $x_i = 3$ $(i = 1, 2, \ldots, 8)$, and each facet is a triangle.

10 We proof by induction on n. When $n = 4$, the number of pairs of points which have the determined distances in M_4 is $\frac{1}{2} \cdot 4(4 - 3) + 4 = 6$. But 4 points just form $\binom{4}{2} = 6$ pairs of points, so M_4 is stable. Suppose that when $n = k$ $(k \geqslant 4)$, the conclusion holds. Then when $n = k + 1$, assume that the number of pairs of points which have the determined distances in M_{k+1} is $\frac{1}{2}(k + 1)(k - 2) + 4$, then the sum of pairs of points which have the determined distances meeting every point in M_{k+1} equals $(k + 1)(k - 2) + 8$. By the mean value principle, there exists a point A such that the number of pairs of points which have the determined distances meeting point A is $l \leqslant \frac{1}{k + 1}[(k + 1) (k - 2) + 8] = k - 1 + \frac{7 - k}{k + 1}$. Since $\frac{7 - k}{k + 1} < 1$, $l \leqslant k - 1$. Deleting the point A, we know among the remaining figure, there are k points and the number of pairs of points which have the determined distances is at least $\frac{1}{2}(k + 1)(k - 2) + 4 - (k - 1) = \frac{1}{2}k(k - 3) + 4$. Applying the inductive hypothesis, we obtain the set consisting of the k points is stable and there are at least $\frac{1}{2}(k + 1)(k - 2) + 4 - \binom{k}{2} = 3$ points: B, C and D of the k points such that the distances between A and each of B, C and D are determined. Assume that $AB = x$, $AC = y$, and $AD = z$ $(x, y, z$ are determinate real numbers), then A is unique. In fact, if there is another point $A' \neq A$ such that $A'B = x$, $A'C = y$, and $A'D = z$, then the three points B, C and D are on the perpendicular bisector of AA', and it contradicts the given conditions. Thus we complete the proof.

Exercise 3

1 (1) Let $f(x) = a_0 + a_1x + a_2x^2 + \cdots + a_nx^n + \cdots,$ ①

then

$$-3xf(x) = -3a_0x - 3a_1x^2 - 3a_2x^3 - \cdots - 3a_{n-1}x^n - \cdots,$$ ②

$$2x^2f(x) = 2a_0x^2 + 2a_1x^3 + 2a_2x^4 + \cdots + 2a_{n-2}x^n + \cdots.$$ ③

With ① + ② + ③ and the given recurrence relation, we obtain $(1 - 3x + 2x^2)f(x) = 2 - x$, so

$$f(x) = \frac{2-x}{1-3x+2x^2} = \frac{2-x}{(1-x)(1-2x)}$$

$$= \frac{3}{1-2x} - \frac{1}{1-x} = 3\sum_{n=0}^{\infty}(2x)^n - \sum_{n=0}^{\infty}x^n$$

$$= \sum_{n=0}^{\infty}(3 \cdot 2^n - 1)x^n.$$

Hence $a_n = 3 \cdot 2^n - 1$.

(2) Let $f(x) = a_0 + a_1x + a_2x^2 + \cdots + a_nx^n + \cdots,$ ①

then

$$-xf(x) = -a_0x - a_1x^2 - a_2x^3 - \cdots - a_{n-1}x^n - \cdots,$$ ②

$$-6x^2f(x) = -6a_0x^2 - 6a_1x^3 - 6a_2x^4 - \cdots - 6a_{n-2}x^n + \cdots,$$ ③

$$\frac{12}{1-x} = 12 + 12x + 12x^2 + \cdots + 12x^n + \cdots.$$ ④

With ① + ② + ③ + ④ and the given recurrence relation, we obtain

$$(1 - x - 6x^2)f(x) + \frac{12}{1-x} = 16 + 11x,$$

so $$f(x) = \frac{4 - 5x - 11x^2}{(1 - x - 6x^2)(1-x)}$$

$$= \frac{4 - 5x - 11x^2}{(1-x)(1+2x)(1-3x)}$$

$$= \frac{A}{1-x} + \frac{B}{1+2x} + \frac{C}{1-3x}.$$

Then $A = \dfrac{4-5x-11x^2}{(1+2x)(1-3x)}\Big|_{x=1} = 2$, $B = \dfrac{4-5x-11x^2}{(1-x)(1-3x)}\Big|_{x=-\frac{1}{2}} = 1$,

$C = \dfrac{4-5x-11x^2}{(1-x)(1+2x)}\Big|_{x=\frac{1}{3}} = 1$,

so

$$f(x) = \frac{2}{1-x} + \frac{1}{1+2x} + \frac{1}{1-3x}$$

$$= 2\sum_{n=0}^{\infty} x^n + \sum_{n=0}^{\infty} (-2x)^n + \sum_{n=0}^{\infty} (3x)^n$$

$$= \sum_{n=0}^{\infty} (2 + (-2)^n + 3^n) x^n.$$

Hence $a_n = 3^n + (-2)^n + 2$.

2 (1) On one hand, the coefficient of x^n in $(1+x)^{2n} = \sum_{k=0}^{2n} \binom{2n}{k} x^k$ is $\binom{2n}{n}$. On the other hand, the coefficient of x^n in

$$(1+x)^{2n} = (1+x)^n (1+x)^n = \left(\sum_{i=0}^{n} \binom{n}{i} x^i\right)\left(\sum_{j=0}^{n} \binom{n}{j} x^j\right)$$

is $\sum_{i=0}^{n} \binom{n}{i}\binom{n}{n-i} = \sum_{i=0}^{n} \binom{n}{i}^2$.

So $\sum_{i=0}^{n} \binom{n}{i}^2 = \binom{2n}{n}$.

(2) On one hand, the coefficient of x^{n+1} in $(1+x)^{2n} = \sum_{k=0}^{2n} \binom{2n}{k} x^k$ is $\binom{2n}{n+1}$. On the other hand, the coefficient of x^{n+1} in

$$(1+x)^{2n} = (1+x)^n (1+x)^n = \left(\sum_{i=0}^{n} \binom{n}{i} x^i\right)\left(\sum_{j=0}^{n} \binom{n}{j} x^j\right)$$

is $\sum_{k=1}^{n} \binom{n}{k}\binom{n}{n+1-k}$. So

$$\sum_{k=1}^{n} \binom{n}{k}\binom{n}{n+1-k} = \binom{2n}{n+1}.$$

(3) On one hand, the coefficient of x^n in

$$(1+x)^{n+1} = \sum_{k=0}^{n+1} \binom{n+1}{k} x^k$$

is $\binom{n+1}{n} = n+1$. On the other hand, the coefficient of x^n in

$$(1+x)^{n+1} = \frac{(1-x^2)^n}{(1-x)^n} = \left(\sum_{k=0}^{n}(-1)^n \binom{n}{k} x^{2k}\right)\left(\sum_{j=0}^{\infty}\binom{n+j}{n}x^j\right)$$

is

$$\sum_{k=0}^{[n/2]}(-1)^k \binom{n}{k}\binom{n+(n-2k)}{n} = \sum_{k=0}^{[n/2]}(-1)^n \binom{n}{k}\binom{n}{n+1-k}.$$

So

$$\sum_{k=0}^{[n/2]}(-1)^k \binom{n}{k}\binom{2n-2k}{n} = \binom{2n}{n+1}.$$

(4) Note that $\binom{n+k+1}{2k+1} = \binom{n+k+1}{n-k}$ and the coefficient of x^{n-k} in

$$(2-x)^{n+k+1} = \sum_{i=0}^{n+k+1}(-1)^i \binom{n+k+1}{i} \cdot 2^{n+k+1-i}x^i$$

is

$$(-1)^{n-k}\binom{n+k-1}{n-k} \cdot 2^{n+k+1-(n-k)} = (-1)^{n-k}2^{2k+1}\binom{n+k+1}{n-k},$$

so the coefficient of x^n in $f(x) = \sum_{k=0}^{n}(-x+2)^{n+k+1} \cdot x^k$ is

$$A_n = \sum_{k=0}^{n}(-1)^{n-k}2^{2k+1}\binom{n+k+1}{n-k}.$$

On the other hand, in the equality

$$
\begin{aligned}
f(x) &= (2-x)^{n+1}(1-x)^{-2}\\
&= [1+(1-x)]^{n+1}(1-x)^{-2}\\
&= \binom{n+1}{0}(1-x)^{n-1} + \cdots + \binom{n+1}{1}(1-x)^{n-2} + \cdots\\
&\quad + \binom{n+1}{n}(1-x)^{-1} + \binom{n+1}{n+1}(1-x)^{-2}.
\end{aligned}
$$

Only last two terms contain the term x^n and the sum of coefficients of x^n of the two terms is

$$\binom{n+1}{n} + \binom{n+1}{1} = 2(n+1).$$

So

$$\sum_{k=0}^{n} (-1)^{n-k} 2^{2k} \binom{n+k+1}{2k+1} = (n+1).$$

(5) When $k \geqslant p$, the coefficient of x^p in $(1+x)^k = \sum_{i=0}^{k} \binom{n}{i} x^i$ is $\binom{k}{p}$ and when $k < p$, the coefficient of x^p in $(1+x)^k = \sum_{i=0}^{k} \binom{n}{i} x^i$ is zero, so the coefficient of x^p in

$$f(x) = \sum_{k=0}^{n} (-1)^k \binom{n}{k} (1+x)^k$$

is $\sum_{k=p}^{n} (-1)^k \binom{n}{k} \binom{k}{p}$. On the other hand, the coefficient of x^p in

$$f(x) = [1 - (1+x)]^n = (-1)^n x^n$$

is

$$(-1)^n \delta_{p,n} = \begin{cases} 0, & p \neq n, \\ (-1)^n, & p = n. \end{cases}$$

So

$$\sum_{k=p}^{n} (-1)^k \binom{n}{k} \cdot \binom{k}{p} = (-1)^n \delta_{p,n}.$$

Exercise 4

1 Two sides are divided by $\sqrt{a_{n-1} a_{n-2}}$ in (1), we obtain $\sqrt{\dfrac{a_n}{a_{n-1}}} =$

$1+2\sqrt{\dfrac{a_{n-1}}{a_{n-2}}}$. Set $b_n = \sqrt{\dfrac{a_n}{a_{n-1}}}$, then $b_n = 1 + 2b_{n-1}$, i.e. $b_{n+1} + 1 = 2(b_{n-1} +$

1). So $b_n + 1 = (b_1 + 1) \cdot 2^{n-1} = 2^n$. This is $\dfrac{a_n}{a_{n-1}} = (2^n - 1)^2$, so

$$a_n = \frac{a_n}{a_{n-1}} \cdot \frac{a_{n-1}}{a_{n-2}} \cdot \cdots \cdot \frac{a_2}{a_1} \cdot a_1 = \prod_{k=1}^{n} (2^k - 1)^2.$$

2 Both sides are divided by 2^{n+1}, we obtain that $\dfrac{a_{n+1}}{2^{n+1}} = -\dfrac{3}{2} \cdot$

$\dfrac{a_n}{2^n} + \dfrac{1}{2}$. Set $b_n = \dfrac{a_n}{2^n}$, then $b_{n+1} = -\dfrac{3}{2}b_n + \dfrac{1}{2}$, i.e. $b_{n+1} - \dfrac{1}{5} =$

$-\dfrac{3}{2}\left(b_n - \dfrac{1}{5}\right)$ (Note that $\dfrac{1}{5}$ is the root of equation $x = -\dfrac{3}{2}x +$

$\dfrac{1}{2}$). Thus

$$b_n - \frac{1}{5} = \left(b_0 - \frac{1}{5}\right)\left(-\frac{3}{2}\right)^n = \left(a_0 - \frac{1}{5}\right)\left(-\frac{3}{2}\right)^n,$$

and

$$a_n = 2^n b_n = 2^n\left[\left(a_0 - \frac{1}{5}\right)\left(-\frac{3}{2}\right)^n + \frac{1}{5}\right]$$

$$= (-3)^n\left[\left(a_0 - \frac{1}{5}\right) + \frac{1}{5}\left(-\frac{2}{3}\right)^n\right].$$

Since $\left(-\dfrac{2}{3}\right)^n \to 0$ $(n \to \infty)$, if $a_0 - \dfrac{1}{5} \neq 0$, and n is great enough,

then a_n and $(-3)^n\left(a_0 - \dfrac{1}{5}\right)$ have the same sign. Hence $\{a_n\}$ is

increasing if and only if $a_0 = \dfrac{1}{5}$.

3 Let $S_n = a_1 + a_2 + \cdots + a_n$, then

$$a_n = S_n - S_{n-1} = n^2 a_n - (n-1)^2 a_{n-1},$$

i.e. $a_n = \dfrac{n-1}{n+1}a_{n-1}$. So

$$a_n = \frac{a_n}{a_{n-1}} \cdot \frac{a_{n-1}}{a_{n-2}} \cdot \dots \cdot \frac{a_2}{a_1} \cdot a_1 = \frac{n-1}{n+1} \cdot \frac{n-2}{n} \cdot \frac{n-3}{n-1} \cdot \dots \cdot \frac{1}{3} \cdot \frac{1}{2}$$

$$= \frac{1}{n(n+1)}.$$

4   The characteristic equation is $r^2 - 2r - 1 = 0$, and the characteristic roots are $r_{1,2} = 1 \pm \sqrt{2}$. So $a_n = c_1(1+\sqrt{2})^n + c_2(1 - \sqrt{2})^n$. From $a_0 = 0$, and $a_1 = 1$, we know that $c_1 = \frac{1}{2\sqrt{2}}$, $c_2 = -\frac{1}{2\sqrt{2}}$. Thus

$$a_n = \frac{1}{2\sqrt{2}}(1+\sqrt{2})^n - \frac{1}{2\sqrt{2}}(1-\sqrt{2})^n.$$

Set $(1+\sqrt{2})^n = A_n + B_n \sqrt{2}$ (A_n, $B_n \in \mathbf{N}_+$), then $(1-\sqrt{2})^n = A_n - B_n \sqrt{2}$, $a_n = B_n$, and $A_n^2 - 2B_n^2 = (-1)^n$. It implies A_n is odd. Let $n = 2^k(2t+1)$ (k, t are nonnegative integers). Thus we need to prove that B_n is a multiple of 2^k but not 2^{k+1}. We prove by induction on n. When $k = 0$, then $n = 2t+1$ is odd, and A_n is also odd, so $2B_n^2 = A_n^2 + 1 \equiv 2 \pmod 4$. Thus B_n is odd, and it means that B_n is a multiple of 2^0 but not 2^1. Assume that when $k = m$, B_n is a multiple of 2^m but not 2^{m+1}, and when $k = m+1$, with $(A_n + B_n \sqrt{2})^2 = (1+\sqrt{2})^{2n} = A_{2n} + \sqrt{2}B_{2n}$, we obtain $B_{2n} = 2A_nB_{2n}$ and since A_n is odd, B_{2m} is a multiple of 2^{m+1} but not 2^{m+2}, and $2n = 2^{k+1}(2t+1)$. Therefore we complete the proof.

5   Assume that there are a_n distinct ways. Obviously, $a_1 = 1$, $a_2 = 2$. When the chessboard is $2 \times n$, if we use 1×2 rectangles to cover two small squares on the leftmost column, then there are a_{n-1} distinct ways to cover the remaining $2 \times (n-1)$ rectangles on the chessboard, and if we use two 1×2 rectangles to cover the leftmost 2×2 square on the chessboard, then there are a_{n-2} distinct ways to cover the remaining $2 \times (n-2)$ rectangles on the chessboard. Thus $a_n = a_{n-1} + a_{n-2}$. From example 4, we obtain

$$a_n = \frac{1}{\sqrt{5}}\left[\left(\frac{1+\sqrt{5}}{2}\right)^{n+1} - \left(\frac{1-\sqrt{5}}{2}\right)^{n+1}\right].$$

6 Obviously, $a_1 = 2$. Since the sphere is divided into a_{n-1} regions by $n-1$ big circles, and no three of circles are concurrent, the number of intersecting points of the nth great circle and the $n-1$ big circles is $2(n-1)$ and the nth big circle is divided into $2(n-1)$ arcs by these intersecting points. Hence each original region is divided into two regions by these arcs, and the number of incremental regions is $2(n-1)$, and this is $a_n = a_{n-1} + 2(n-1)$. Therefore

$$a_n = a_1 + \sum_{k=2}^{n}(a_k - a_{k-1}) = 2 + 2\sum_{k=2}^{n}(k-1)$$

$$= 2 + 2 \cdot \frac{n(n-1)}{2} = n^2 - n + 2.$$

7 Obviously, $a_1 = 3$. Among the n-digit numbers satisfying the given conditions, the number of n-digit numbers with leading digit 1 is $4^{n-1} - a_{n-1}$ and the number of n-digit numbers with leading digit $k = 2$, 3 or 4 is $3a_{n-1}$, thus

$$a_n = 4^{n-1} - a_{n-1} + 3a_{n-1} = 2a_{n-1} + 4^{n-1}.$$

Both sides are divided by 4^n: $\frac{a_n}{4^n} = \frac{1}{2} \cdot \frac{a_{n-1}}{4^{n-1}} + \frac{1}{4}$, and set $b_n = \frac{a_n}{4^n}$,

then $b_n = \frac{1}{2}b_{n-1} + \frac{1}{4}$, i. e. $b_n - \frac{1}{2} = \frac{1}{2}\left(b_{n-1} - \frac{1}{2}\right)$.

So

$$b_n - \frac{1}{2} = \left(b_1 - \frac{1}{2}\right)\left(\frac{1}{2}\right)^{n-1} = \left(\frac{3}{4} - \frac{1}{2}\right)\left(\frac{1}{2}\right)^{n-1} = \frac{1}{2^{n+1}}.$$

Therefore

$$a_n = 4^n b_n = 4^n\left(\frac{1}{2^{n+1}} + \frac{1}{2}\right) = \frac{1}{2}(2^n + 4^n).$$

8 Assume that the ball returns to A after n times in a_n distinct ways. It can be easily concluded that $a_1 = 0$, and $a_2 = 3$ and there are 3^{n-1} distinct ways to pass the ball $n-1$ times continually. These ways are divided into two classes: (1) the ball returns to A after the $(n-1)^{\text{th}}$ times in a_{n-1} distinct ways and (2) the ball is passed to one of B, C, D after the $(n-1)^{\text{th}}$ times but the ball returns to A after the nth time in a_n distinct ways. Hence $a_{n-1} + a_n = 3^{n-1}$, i.e. $\dfrac{a_n}{3^n} + \dfrac{1}{3} \cdot \dfrac{a_{n-1}}{3^{n-1}} = \dfrac{1}{3}$. Set $b_n = \dfrac{a_n}{3^n}$, thus $b_n + \dfrac{1}{3} b_{n-1} = \dfrac{1}{3}$, i.e. $b_n - \dfrac{1}{4} = -\dfrac{1}{3} \cdot \left(b_{n-1} - \dfrac{1}{4} \right)$.

$\left(\text{Note that } \dfrac{1}{4} \text{ is the root of the equation } x + \dfrac{1}{3}x = \dfrac{1}{3}.\right)$ Hence $b_n - \dfrac{1}{4} = \left(b_1 - \dfrac{1}{4} \right) \cdot \left(-\dfrac{1}{3} \right)^{n-1} = \left(-\dfrac{1}{4} \right)\left(-\dfrac{1}{3} \right)^{n-1}$, and

$$a_n = 3^n b_n = 3^n \left[\left(-\dfrac{1}{4} \right)\left(-\dfrac{1}{3} \right)^{n-1} + \dfrac{1}{4} \right] = \dfrac{1}{4}(3^n + (-1)^n \cdot 3).$$

Especially, $a_{10} = \dfrac{1}{4}(3^{10} + (-1)^{10} \cdot 3) = 14\,763$.

9 Let $a_n = (1+\sqrt{3})^{2n+1} + (1-\sqrt{3})^{2n+1} = (1+\sqrt{3})(4+2\sqrt{3})^n + (1-\sqrt{3})(4-2\sqrt{3})^n$, and

$$b_n = \dfrac{a_n}{2^{n+1}} = \dfrac{1+\sqrt{3}}{2}(2+\sqrt{3})^n + \dfrac{1-\sqrt{3}}{2}(2-\sqrt{3})^n.$$

Thus the characteristic roots corresponding to $\{b_n\}$ are $2 \pm \sqrt{3}$, and the characteristic equation is $[r - (2+\sqrt{3})][r - (2-\sqrt{3})] = 0$, i.e. $r^2 - 4r + 1 = 0$. So $\{b_n\}$ satisfies the recurrence relation: $b_n = 4b_{n-1} - b_{n-2}$ $(n \geqslant 2)$, with $b_0 = \dfrac{1+\sqrt{3}}{2} + \dfrac{1-\sqrt{3}}{2} = 1$, and

$$b_1 = \dfrac{1+\sqrt{3}}{2}(2+\sqrt{3}) + \dfrac{1-\sqrt{3}}{2}(2-\sqrt{3}) = 5.$$

Since b_0, b_1 are positive integers, assume that b_{n-2}, b_{n-1} $(n \geqslant 2)$ are positive integers, then $b_n = 4b_{n-1} - b_{n-2}$ is also a positive integer. For any nonnegative integer n, b_n is a positive integer. We obtain for any nonnegative integer n, a_n is divisible by 2^{n+1}. Since $0 < (1-\sqrt{3})^{2n+1} < 1$, $a_n = (1+\sqrt{3})^{2n+1} + (1-\sqrt{3})^{2n+1} = [(1+\sqrt{3})^{2n+1}]$. So $[(1+$

$\sqrt{3}\,)^{2n+1}]$ is divisible by 2^{n+1}.

10 Assume that $x_n = \sum\limits_{k=0}^{n} \binom{2n+1}{2k+1} 2^{3k} = \dfrac{1}{\sqrt{8}} \sum\limits_{k=0}^{n} \binom{2n+1}{2k+1} (\sqrt{8}\,)^{2k+1}$,

and $y_n = \dfrac{1}{\sqrt{8}} \sum\limits_{k=0}^{n} \binom{2n+1}{2k} (\sqrt{8}\,)^{2k}$, then $x_n + y_n = \dfrac{1}{\sqrt{8}} (\sqrt{8}+1)^{2n+1}$, and

$x_n - y_n = \dfrac{1}{\sqrt{8}} (\sqrt{8}-1)^{2n+1}$. So

$$x_n = \frac{1}{4\sqrt{2}} [(\sqrt{8}+1)^{2n+1} + (\sqrt{8}-1)^{2n+1}]$$

$$= \frac{1}{4\sqrt{2}} [(\sqrt{8}+1)(9+4\sqrt{2})^n + (\sqrt{8}-1)(9-4\sqrt{2})^n].$$

Hence the characteristic roots of x_n are $9 \pm 4\sqrt{2}$, and the characteristic equation is

$$[r - (9+4\sqrt{2})][r - (9-4\sqrt{2})] = 0,$$

i.e. $r^2 - 18r + 49 = 0$. Thus $\{x_n\}$ satisfies the recurrence relation:

$$x_n = 18x_{n-1} - 49x_{n-2}, \qquad\qquad ①$$

and $\qquad\qquad 2x_{n-1} = 36x_{n-2} - 98x_{n-3}. \qquad\qquad ②$

① $-$ ②: $x_n = 20x_{n-1} - 85x_{n-2} + 98x_{n-3} \equiv 3x_{n-3} \pmod 5$.
Hence x_n is divisible by 5 if and only if x_{n-3} is divisible by 5. Since
each of $x_0 = 1$, $x_1 = \binom{3}{1} \cdot 2^0 + \binom{3}{3} \cdot 2^3 = 11$, and

$$x_2 = \binom{5}{1} \cdot 2^0 + \binom{5}{3} \cdot 2^3 + \binom{5}{5} \cdot 2^5 = 149$$

is not divisible by 5, hence for any nonnegative integer n, $x_n \equiv 3x_{n-3}$ $\pmod 5$ is not divisible by 5. Next,

$$x_n = 18x_{n-1} - 49x_{n-2} \equiv 4x_{n-1} \pmod 7.$$

Hence x_n is divisible by 7 if and only if x_{n-1} is divisible by 7. Since $x_0 = 1$ is not divisible by 7, $x_n \equiv 4x_{n-1} \pmod 7$ is not divisible by 7. As

$(5, 7) = 1$, we know for any nonnegative integer n, x_n is not divisible by 35.

11 (1) By assumption, we obtain that $a_1 = 5$ and $\{a_n\}$ is strictly increasing with

$$2a_{n+1} - 7a_n = \sqrt{45a_n^2 - 36}.$$

Square both sides, and we yield

$$a_{n+1}^2 - 7a_n a_{n+1} + a_n^2 + 9 = 0, \qquad ①$$

$$a_n^2 - 7a_{n-1}a_n + a_{n-1}^2 + 9 = 0. \qquad ②$$

① $-$ ② (since $a_{n+1} - a_n > 0$):

$$a_{n+1} = 7a_n - a_{n-1}. \qquad ③$$

It follows that with $a_0 = 1$, $a_1 = 5$ and ③ a_n is a positive integer for each $n \in \mathbf{N}_+$.

(2) From ①, we obtain

$$(a_{n+1} + a_n)^2 = 9(a_n a_{n+1} - 1),$$

so

$$a_n a_{n+1} - 1 = \left(\frac{a_{n+1} + a_n}{3}\right)^2.$$

By (1), a_n and a_{n+1} are positive integers and $\dfrac{a_n + a_{n+1}}{3}$ is a rational number. Since $\left(\dfrac{a_{n+1} + a_n}{3}\right)^2 = a_n a_{n+1} - 1$ is a positive integer, and so is $\dfrac{a_n + a_{n+1}}{3}$. Thus $a_n a_{n+1} - 1$ is the square of an integer.

Exercise 5

1 When there are exactly two distinct digits in $\overline{abcd}$, the number of ways of taking two distinct digits from 4 distinct digits is $\binom{4}{2}$, and the less digit must be the first digit or the third digit, while the bigger number must be the second digit or the fourth digit, i. e. the 4-digit number is determined uniquely by the two digits. Hence in this case, the number of distinct 4-digit numbers is $\binom{4}{2} = 6$. When there are exactly three distinct digits in $\overline{abcd}$, the number of ways of taking three distinct digits from 4 distinct digits is $\binom{4}{3}$, and the less digit must be the first digit. If the first digit and third digit are the same, then P_2^2 4-digit numbers are formed by the three digits. If the second digit and the fourth digit are the same, then P_2^2 4-digit numbers are formed by the three digits. Hence in this case, the number of distinct 4-digit numbers is $\binom{4}{3}(P_2^2 + P_2^2) = 16$. When there are exactly four distinct digits in $\overline{abcd}$, the less digit must be the first digit, the other three digits could be arranged in P_3^3 ways. Hence in this case, the number of distinct 4-digit numbers is $P_3^3 = 6$. Summing up the above, the required number of distinct 4-digit numbers is $6 + 16 + 6 = 28$.

2 Let S_n denote the set of the n-digit numbers consisting of 1, 2, 3, 4, 5, and 6 in which the digits 1 and 6 are nonadjacent. Let $A_n \subseteq S_n$ denote the set of n-digit numbers in which there are at least three distinct digits and $B_n \subseteq S_n$ denote the set of n-digit numbers in which there are at most two distinct digits. Set $s_n = |S_n|$, $a_n = |A_n|$, and $b_n = |B_n|$, then $a_n = s_n - b_n$ and we obtain $s_1 = 6$, $s_2 = 6^2 - 2 = 34$ and

$$b_5 = 6 + \left(\binom{6}{2} - 1\right)(2^5 - 2) = 426.$$

Let $X_n \subseteq S_n$ denote the set of n-digit numbers with the leading digit 1 or 6 and let $Y_n \subseteq S_n$ denote the set of n-digit numbers by leading digit 2, 3, 4 or 5. Set $x_n = |X_n|$, $y_n = |Y_n|$, then we obtain the following three recurrence relations:

$$s_n = x_n + y_n, \qquad\qquad ①$$
$$x_n = x_{n-1} + 2y_{n-1}, \qquad\qquad ②$$

and

$$y_n = 4s_{n-1}. \qquad \qquad ③$$

Substituting ③ to ① and ②, we obtain

$$s_n = x_n + 4s_{n-1}, \qquad \qquad ④$$

and

$$x_n = x_{n-1} + 8s_{n-2}. \qquad \qquad ⑤$$

From ④, we know

$$x_n = s_n - 4s_{n-1}. \qquad \qquad ⑥$$

Substituting ⑥ to ⑤, we obtain $s_n = 5s_{n-1} + 4s_{n-2}$. Combining $s_1 = 6$, $s_2 = 34$, we could obtain $s_3 = 194$, $s_4 = 1106$, and $s_5 = 6306$. Therefore the required number of 5-digit numbers is

$$a_5 = s_5 - b_5 = 6306 - 426 = 5880.$$

3 The number of 4-digit numbers with thousands digit 4 or 6 is $2 \times 4 \times 8 \times 7 = 448$ and the number of 4-digit numbers with thousands digit 5 is $1 \times 5 \times 8 \times 7 = 280$. Therefore the required number of evens with 4-digit is $448 + 280 = 728$.

4 Let the lengths of n line segments be a_1, a_2, ..., a_n respectively. Firstly, if there exists some $a_k > 1$, then 1, 2, and a_k are lengths of the sides of a triangle. Secondly, assume that $a_1 \leqslant a_2 \leqslant \cdots \leqslant a_n \leqslant 1$, if $a_{n-1} > \frac{1}{2}$, then $a_{n-1} + a_n > 1$, thus 1, a_{n-1}, a_n are the lengths of the sides of a triangle. If $a_{n-1} \leqslant \frac{1}{2}$ and there exists k such that $a_k + a_{k+1} > a_{k+2}$, then a_k, a_{k+1}, a_{k+2} are the lengths of the sides of a triangle. If $a_{n-1} \leqslant \frac{1}{2}$ and for any k, the inequality $a_k + a_{k+1} \leqslant a_{k+2}$ holds, then $a_k \leqslant \frac{1}{2} a_{k+2}$. Thus when n is odd,

$$a_1 + a_2 + \cdots + a_n \leqslant 2(a_2 + a_4 + \cdots + a_{n-1}) + a_n$$
$$\leqslant 2 \left(\frac{1}{2} + \frac{1}{4} + \cdots + \frac{1}{2^{(n-1)/2}} \right) + 1 < 3.$$

When n is even,

$$a_1 + a_2 + \cdots + a_n$$
$$\leqslant 2(a_1 + a_3 + \cdots + a_{n-1}) + a_n$$
$$\leqslant 2\left(\frac{1}{2} + \frac{1}{4} + \cdots + \frac{1}{2^{n/2}}\right) + 1$$
$$< 3,$$

and this is a contradiction. This completes the proof.

5 Since the number of subsets of S is 2^n, then the number of ways to select subsets X and Y orderly is $P_2^{2^n} = 2^n(2^n - 1)$. When $X \subset Y(X \neq Y)$, and set $|Y| = i$ $(1 \leqslant i \leqslant n)$, then the number of ways to select Y is $\binom{n}{i}$. Since X is the proper subset of Y, then there are $2^i - 1$ ways to select X, and the number of ways to select $X \subset Y(X \neq Y)$ is

$$\sum_{i=0}^{n} \binom{n}{i}(2^i - 1) = 3^n - 2^n.$$

Similarly, the number of ways to select $Y \subset X(Y \neq X)$ is

$$\sum_{i=0}^{n} \binom{n}{i}(2^i - 1) = 3^n - 2^n.$$

Therefore we yield that required number of ways to select X and Y orderly is

$$2^n(2^n - 1) - 2(3^n - 2^n) = 2^{2n} - 2 \cdot 3^n + 2^n.$$

6 We know that among any five integers, there are three numbers whose sum is a multiple of 3 and the coordinates of the barycenter of $\triangle A_i A_j A_k$ is $\left(\frac{x_1 + x_2 + x_3}{3}, \frac{y_1 + y_2 + y_3}{3}\right)$.

(1) If among the abscissas of all A_i there are five numbers which have the same remainders with module 3, then without loss of generality, assume that $x_1 \equiv x_2 \equiv x_3 \equiv x_4 \equiv x_5 \pmod{3}$. Since among $y_1, y_2, y_3, y_4,$ and y_5 there are three numbers, say y_i, y_j, y_k $(1 \leqslant$

$i < j < k \leqslant 5)$, whose sum is a multiple of 3. Thus the barycenter of $\triangle A_i A_j A_k$ is an integral point.

(2) Similarly, if among the ordinate of A_i there are five numbers which have the same remainders with module 3, then the conclusion holds too.

(3) If all the remainders with module 3 of any five abscissas of A_i are not the same, and so are the ordinates. Then, the remainders of x_i and y_i $(i = 1, 2, \ldots, 9)$ range over 0, 1 and 2 respectively, so there are at least two remainders of x_i (or y_i) appear 3 times. Without loss of generality, assume that

$$x_1 \equiv x_2 \equiv x_3 \equiv 0 \; (\text{mod} \, 3),$$
$$x_4 \equiv x_5 \equiv x_6 \equiv 1 \; (\text{mod} \, 3).$$

If $y_1 \equiv y_2 \equiv y_3 \, (\text{mod} \, 3)$ or $y_4 \equiv y_5 \equiv y_6 \, (\text{mod} \, 3)$, then the conclusion holds. Otherwise, with module 3, the remainders of y_1, y_2 and y_3 take at least two distinct values α, β ($\alpha \neq \beta$ and $\alpha, \beta \in \{0, 1, 2\}$) and we set $\{\alpha, \beta, \gamma\} = \{0, 1, 2\}$. Similarly, with module 3, the remainders of y_4, y_5 and y_6 take at least two distinct values α and β, or α and γ, or β and γ. In other words, with module 3 the remainders of y_1, y_2, y_3, y_4, y_5 and y_6 are classified two possible cases: (1) These remainder take all values of $\{\alpha, \beta, \gamma\} = \{0, 1, 2\}$; (2) These remainder take values of α and β, but each value takes 2 – 4 times. In this case, we take $k \in \{7, 8, 9\}$ such that $x_k \equiv 2 \, (\text{mod} \, 3)$, then there exist $1 \leqslant i \leqslant 3 < j \leqslant 6$ such that

$$x_i + x_j + x_k \equiv 0 + 1 + 2 \equiv 0 \; (\text{mod} \, 3)$$

and

$$y_i + y_j + y_k \equiv \alpha + \beta + \gamma \equiv 0 \; (\text{mod} \, 3),$$

or

$$y_i + y_j + y_k \equiv \alpha + \alpha + \alpha \equiv 0 \; (\text{mod} \, 3)$$

or

$$y_k \equiv \beta + \beta + \beta \equiv 0 \; (\text{mod} \, 3).$$

Therefore the barycenter of $\triangle A_i A_j A_k$ is an integral point. This completes the proof.

7 We use n points (no four of which are coplanar) to represent n persons. If two persons are mutually unacquainted then the two corresponding points are connected by a red line segment. Otherwise, points are connected by a blue line segment. Thus we obtain a 2-colored complete graph K_n in which there is not a red triangle and the problem is equivalent to find the smallest positive integer n such that there exists a blue complete graph K_4 in K_n, i. e. there exist four points A, B, C, D in K_n such that each pair of points is connected a blue line segment. Firstly, as figure, in a 2-colored complete graph K_8 (the solid lines is colored red and the dashed lines is colored blue), there is neither red triangle nor blue complete graph K_4. If we delete some points and all line segments meeting these points in this graph, and in the remaining graph, there is neither a red triangle nor blue complete graph K_4. So the required smallest positive integer $n \geqslant 9$. When $n = 9$, we consider the following three cases:

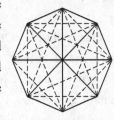

(1) There exists a point A_1 such that there are 4 red line segments, say $A_1 A_2$, $A_1 A_3$, $A_1 A_4$, $A_1 A_5$, meeting A_1. With the given conditions, we know, the line segments connecting each pair of four points A_2, A_3, A_4, A_5 are colored blue, thus the conclusion holds.

(2) There exists a point A_1 such that there are at most 2 red line segments meeting A_1, i. e. there are at least 6 blue line segments, say $A_1 A_2$, $A_1 A_3$, $A_1 A_4$, $A_1 A_5$, $A_1 A_6$, and $A_1 A_7$, metting A_1. Considering the complete graph K_6 with 6 vertices A_2, A_3, A_4, A_5, A_6, A_7, and Ramsey's theorem (Example 1 in Chapter 2), we obtain that there exists a red or a blue triangle in this complete graph K_6. But from the given conditions, we know there is not a red triangle in the graph, hence there must exist a blue triangle, say $\triangle A_2 A_3 A_4$, in the graph, and there exists a blue complete graph K_4 with four vertices:

A_1, A_2, A_3, A_4. Hence in this case, the conclusion also holds;

(3) The numbers of red line segments meeting each of points are all 3, thus the total of red line segments in the graph is $\dfrac{3 \times 9}{2} = \dfrac{27}{2} \neq a$ positive integer, and it is a contradiction, i. e. the case (3) is impossible.

This completes the proof.

8 As shown in the figure, there are 9 line segments connecting 7 points, such that the condition is satisfied, hence the required smallest number of line segments $n \leqslant 9$. If there are 8 line segments connecting 7 points, then we prove that there exist three points such that no line segments connect the three points. Since there are 16 end points in the 8 line segments, by the mean value principle, there exists a point A such that there are at most $\left[\dfrac{16}{8}\right] = 2$ line segments meeting A.

(1) If there is at most one line segment meeting A, then there are at least other 5 points disjoint with A. Thus there exist at least $\dbinom{5}{2} - 8 = 2$ disjoint pairs of the 5 points. Assume that two points C and D are disjoint, then there is not a line segment connecting three point A , B and C.

(2) If there are two line segments AB and AC meeting A.

(a) If there are 6 line segments connecting other 4 points D, E, F, G, then there is not a line segment connecting three B, C and D.

(b) There are at most 5 line segments joining other 4 points D, E, F, G, then there exist at least $\dbinom{4}{2} - 5 = 1$ disjoint pair of other 4 points D, E, F, G. Assume that two points D and E are disjoint, then there is not a line segment joining the three points A, D and E.

Summing up the above, we obtain that the requires smallest number of line segments is 9.

9 Let the intersection points of the diagonal
lines A_1A_4 and A_2A_5, A_2A_5 and A_3A_6, A_3A_6 and
A_1A_4 be M_1, M_2, M_3 respectively. Firstly,
among the six triangles: $\triangle A_1A_2M_2$, $\triangle A_2A_3M_1$,
$\triangle A_3A_4M_1$, $\triangle A_4A_5M_3$, $\triangle A_5A_6M_3$ and $\triangle A_6A_1M_2$,
there exists a triangle, say $\triangle A_1A_2M_2$, whose

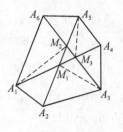

area is less than or equal to $\dfrac{1}{6}S$. We prove this by the mean value
principle and the following fact: the sum of the areas of this six
triangles does not exceed the area S of hexagon $A_1A_2A_3A_4A_5A_6$.
Therefore we obtain that

$$\min\{S_{\triangle A_1A_2A_6}, S_{\triangle A_1A_2A_3}\} \leqslant S_{\triangle A_1A_2M_2} \leqslant \frac{1}{6}S.$$

Exercise 6

1 (1) Consider the following 91 number pairs: $\{i, i+9\}, i = 1,$
$2, 3, \ldots, 91$. Thus each element of the set $\{1, 2, 3, \ldots, 9\} \cup \{92,$
$93, \ldots, 100\}$ belongs to exactly one of the above 91 pairs and each of
other 82 numbers belongs to two of the above 91 pairs. Hence 55
numbers belong to at least $(55 - 18) \times 2 = 92$ pairs of the above 91
pairs. By Pigeonhole Principle, there are at least two numbers
belonging to a pair, hence the their difference equals 9. (2) Among
the following 55 numbers: $1, 2, 3, \ldots, 11; 23, 24, 25, \ldots, 33; 45,$
$46, 47, \ldots, 55; 67, 68, 69, \ldots, 77; 89, 90, 91, \ldots, 99$, there are
not two numbers whose difference equals 11.

2 Let $n = 4m + 1$ $(m \in N_+)$. For any permutation $p = \{a_1,$
$a_2, \ldots, a_n\}$ of $\{1, 2, \ldots, n\}$, with the definition of k_p, we know

$$a_1 + a_2 + \cdots + a_{k_p} < a_{k_p+1} + a_{k_p+2} + \cdots + a_n \qquad ①$$

and

$$a_1 + a_2 + \cdots + a_{k_p+1} \geqslant a_{k_p+2} + a_{k_p+3} + \cdots + a_n. \qquad ②$$

But the equality in ② cannot hold, otherwise,

$$2(a_1 + a_2 + \cdots + a_{k_p+1}) = a_1 + a_2 + \cdots + a_n = \frac{1}{2}n(n+1)$$

$$= (4m+1)(2m+1),$$

i.e. an even = an odd, and it is a contradiction. Hence ② became

$$a_1 + a_2 + \cdots + a_{k_p+1} > a_{k_p+2} + a_{k_p+3} + \cdots + a_n. \qquad ③$$

Thus for the permutation $p' = \{a_n, a_{n-1}, \ldots, a_2 a_1\}$ with reverse order of p, from ① and ③, we know $k_{p'} = n - (k_p + 1)$ i.e. $k_{p'} + k_p = n - 1$. Let each permutation p of $\{1, 2, \ldots, n\}$ and the permutation p' with its reverse order be matched into a pair, then the number of these pairs is $\frac{n!}{2}$. Therefore the required sum of $k_p s$ equals $\frac{(n-1) \cdot n!}{2}$.

3 Without loss of generality, assume that the number of red points in M is the largest and is at least $\left[\dfrac{12 \times 12 - 1}{3}\right] + 1 = 48$ by Pigeonhole Principle. Assume that the number of red points with ordinate i is $a_i (i = 1, 2, \ldots, 12)$, then $a_1 + a_2 + \cdots + a_{12} \geqslant 48$. Let every two red points in the ith row be matched into a pair, thus we know there are $\binom{a_i}{2} = \frac{1}{2} a_i(a_i - 1)$ pairs of red points in the ith row. Thus the sum of pairs of red points equals

$$\sum_{i=1}^{12} \binom{a_i}{2} = \frac{1}{2}\left(\sum_{i=1}^{12} a_i^2 - \sum_{i=1}^{12} a_i\right)$$

$$\geqslant \frac{1}{2}\left[\frac{1}{12}\left(\sum_{i=1}^{12} a_i\right)^2 - \sum_{i=1}^{12} a_i\right]$$

$$= \frac{1}{24}\left(\sum_{i=1}^{12} a_i\right)\left(\sum_{i=1}^{12} a_i - 12\right)$$

$$\geqslant \frac{1}{24} \cdot 48(48 - 12) = 72. \qquad ①$$

On the other hand, if there is no rectangle satisfying the

condition, then the projections of all pairs of red points on the x - axis are different, hence the number of all pairs of red points does not exceed the number of pairs consisting of 12 points $(x, 0)$ $(x = 1,$ 2, ..., 12) on the x - axis, i. e. $\dbinom{12}{2} = 66$. It contradicts ①. Thus the conclusion is true.

4 (1) An obvious pairing method is that the two adjacent cells are matched into a pair, i. e. using 1×2 rectangles to cover the chessboard. If mn is an even, then the chessboard could be covered perfectly by $\dfrac{mn}{2}$ 1×2 rectangles. Player A may move the chessman within the same 1×2 rectangle every time, hence A wins eventually. If mn is odd, then the chessboard could be covered perfectly by $\dfrac{mn - 1}{2}$ 1×2 rectangles except the left lower cell. Hence player A just can enter into another adjacent 1×2 rectangle at first step, and player B may move the chessman within the same 1×2 rectangle every time, and B wins eventually.

(2) We prove that no matter what mn is, player A wins eventually. If mn is even, with the same reason above, we know that the conclusion holds. If mn is odd, we also use $\dfrac{mn - 1}{2}$ 1×2 rectangles to cover the chessboard except the left lower cell and every cell is colored black or white such that any two adjacent cells are colored with distinct colors. Thus every time, player A can move the chessman within the same 1×2 rectangle from the black cell into the white cell, and player B could move the chessman from the white cell to the black cell of the adjacent 1×2 rectangle. In this case, there will be no chessman entering the left lower cell. Therefore the player A wins eventually.

5 Starting with the second small square of the $1 \times n$ rectangle, let two small squares with numbers $2k$ and $2k + 1$ be matched into a pair $\left(k = 1, 2, \ldots, \left[\dfrac{n - 1}{2}\right]\right)$. Among the three chessmen in the small squares with numbers $n - 2$, $n - 1$ and n, there are exactly two chessmen in a pair. Firstly, the player A move the third chessman in the square with number n (or $n - 2$) into the square with number 1, if n is an even (or odd), then this chessman cannot be moved again. Afterwards, when the player B moves one chessman of the remaining two chessmen into some square P, the player A always moves another chessman into square Q such that (P, Q) comprises a pair. Thus A wins eventually.

6 Using an analogous method as in Example 11, we could obtain

$$f(n) = 3\binom{n+2}{4}.$$

7 Let all 201-element subsets of S be divided into 5 classes S_0, S_1, S_2, S_3, S_4, where $S_k = \{A \mid |A| = 201, \text{ and } S_{|A|} \equiv k \pmod 5\}$, where $S_{|A|}$ denotes the sum of elements of A }, ($k = 0, 1, 2, 3, 4$). We construct a mapping $f: S_0 \to S_k (1 \leqslant k \leqslant 4)$ as follows: if $A = \{a_1, a_2, \ldots, a_{201}\} \in S_0$, then

$$f(A) = \{a_1 + k, a_2 + k, \ldots, a_{201} + k\}$$

and if some $a_i + k > 1000$, then we use $a_i + k - 1000$ to take the place of $a_i + k$. Thus

$$\sum_{i=1}^{201} (a_i + k) = \sum_{i=1}^{201} a_i + 201k \equiv k \pmod 5,$$

i.e. $f(A) \in S_k$. Obviously, the mapping f is a bijection; so $|S_k| = |S_0|$ ($1 \leqslant k \leqslant 4$). Thus

$$|S_0| = \frac{1}{5}(|S_0| + |S_1| + |S_2| + |S_3| + |S_4|) = \frac{1}{5}\binom{1000}{201},$$

i.e. the number of good subsets is $\dfrac{1}{5}\dbinom{1000}{201}$.

8 First we observe that P must have at least m friends, since for any set of m people, their common friend has at least m friends. Now we prove that P cannot have more than m friends. Proof by contradiction. Let S be the set of P's friends, and $S = \{B_1, B_2, \ldots,$

B_k }. We have by assumption $k \geqslant m + 1$. We claim that for each $(m-1)$ -element subset S_i' of S there exists an unique person $C_i \in S$ who is a common friend of all the people of S_i'. Consider any such subset $S_i' = \{ B_{i_1} , B_{i_2} , \ldots , B_{i_{m-1}} \}$. Adding P to this set gives a set of size m, and thus by the given condition there exists an unique friend C_i who is a common friend of P, B_{i_1}, B_{i_2}, $\ldots$, $B_{i_{m-1}}$. Since by definition S is the set of all friends of P, so $C_i \in S$. Now we claim that for any two distinct $(m-1)$ -element subset S_i' and S_j' of S, the two corresponding persons C_i and C_j are distinct. Assume for a contradiction that this is not the case, that is, there exist S'_i, $S'_j \subset S$ $(S'_i \neq S'_j)$ with $C_i = C_j$. Take any m-element subset of $S_i' \cup S_j'$, then the people in this set have two distinct friends C_i and P, contradicting the given condition. It follows that each $(m-1)$ -element subset S'_i of S corresponds to a unique person $C_i \in S_i'$. Thus f is an injective mapping from $\Re$ to S, where $\Re$ is a family of all $(m-1)$ -element subsets of S. Hence $| \Re | \leqslant | S |$,

i.e. $\dbinom{k}{m-1} \leqslant k$. But $k \geqslant m + 1$ and $m \geqslant 3$, i.e. $2 \leqslant m - 1 \leqslant k - 2$,

so $\dbinom{k}{m-1} \geqslant \dbinom{k}{2} > k$, and this is a contradiction. Therefore the required largest number is m.

9 Setting $S = \{ 1, 2, \ldots, n \}$ and $T = A_1 \cup A_2 \cup \cdots \cup A_n$, we consider all mappings $f : S \to T$ satisfying the following condition: for any $i \in S$, $f(i) \in A_i$ $(i = 1, 2, \ldots, n)$, let M denote the set consisting of such mappings f. Thus $| M | = | A_1 | \cdot | A_2 | \cdot \cdots \cdot | A_n |$. We claim that there is at least an injective mapping in M. If $f \in M$ is not an injective mapping, then there exist i, $j \in S$, $i \neq j$ such that $f(i) = f(j)$. Thus for such f, the $f(i) = f(j)$ takes at most $| A_i \cap A_j |$ distinct values, and $f(k)(k \neq i, j)$ takes at most $| A_k |$ distinct values. Hence for determinated i, $j \in S(i \neq j)$ the number of the mappings satisfying $f(i) = f(j)$ is at most

$$| A_i \cap A_j | \cdot \prod_{k=1, \, k \neq i, \, j}^{n} | A_k | = \frac{| A_i \cap A_j |}{| A_i | \cdot | A_j |} \cdot | M |.$$

Hence the number of mappings in M which are not injective mappings is at most

$$\sum_{1 \leqslant i < j \leqslant n} \frac{|A_i \cap A_j|}{|A_i| \cdot |A_j|} \cdot |M| < |M|.$$

It follows that there is at least an injective mapping $f_0 \in M$. Let $f_0(i) = a_i (i = 1, 2, \ldots, n)$, then $a_i \in A_i (i = 1, 2, \ldots, n)$ and for any $i, j \in S$ $(i \neq j)$, $a_i = f_0(i) \neq f_0(j) = a_j$.

10 We construct a mapping f as follows. Let any term a_i of the sequence correspond to an ordered pair (x_i, y_i), where x_i is the number of terms of the increasing subsequence with leading term a_i and the longest length, and y_i is the number of terms of the decreasing subsequence with leading term a_i and the longest length. Assume that the conclusion is false, then $1 \leqslant x_i \leqslant n$, $1 \leqslant y_i \leqslant m$. Set $X = \{a_1, a_2, \ldots, a_{mn+1}\}$ and

$$Y = \{(x_i, y_i) \mid 1 \leqslant x_i \leqslant n, 1 \leqslant y_i \leqslant m, \text{ and } x_i, y_i \in \mathbf{N}_+\},$$

then $|X| = mn + 1 > mn = |Y|$, so f is not an injective mapping. Hence there are $a_i, a_j \in X (i \neq j)$ such that

$$(x_i, y_i) = f(a_i) = f(a_j) = (x_j, y_j).$$

But $x_i \geqslant x_j + 1$ if $a_i < a_j$ or $y_i \geqslant y_j + 1$ if $a_i > a_j$, and thus, a contradiction. This completes the proof.

11 We use (a) and (b) to denote several consecutive a's and several consecutive b's in the sequence respectively. Thus the sequences satisfying the given condition just have the following two forms: (I) $(a)(b)(a)(b)(a)(b)(a)$, (II) $(b)(a)(b)(a)(b)(a)(b)$. Thus each sequence with form as (I) or (II) contain three ab's and three ba's. Hence with the condition of the problem, we will need to place 5 a's into the (a)'s and 3 b's into the (b)'s. Applying the formula of the repeated combination (Chapter 1, 1. 3), we obtain that the number of permutations satisfying the condition to be

$$\binom{4+5-1}{5}\binom{3+3-1}{3} = \binom{8}{5}\binom{5}{3} = 560$$

in the case (I) and

$$\binom{3+5-1}{5}\binom{4+3-1}{3}=\binom{7}{5}\binom{6}{3}=420$$

in the case (II). The desired answer is $560+420=980$.

12 Without loss of generality, let $a>b>d$. From $a+c=b+d$, we know $c<d$. Consider the ways to select three number $a>b>c$ from $S=\{1,2,\ldots,n\}$ satisfying $a+c-b\neq b$, and the number of ways to select three numbers $a>b>c$ from $S=\{1,2,\ldots,n\}$ is $\binom{n}{3}$, where the number of ways satisfying $a+c-b=b$, i.e. $a+c=2b$ (a and c have same parity) is $2\binom{n/2}{2}$. Let

$$B=\{(a,b,c)\mid n\geqslant a>b>c\geqslant 1,\ a+c-b\neq b\},$$

then $\mid B\mid=\binom{n}{3}-2\binom{n/2}{2}$. For any $(a,b,c)\in B$, set $d=a+c-b$, then 4-element group (a,b,c,d) satisfies the condition of problem. But each 4-element group (a,b,c,d) corresponds to two triples (a,b,c) and (a,d,c) (since $d>c$), hence the required number of ways is

$$\frac{1}{2}\mid B\mid=\frac{1}{2}\left(\binom{n}{3}-2\binom{n/2}{2}\right)=\frac{1}{24}n(n-2)(n-5).$$

13 For any binary sequence of length $n+1$ $y=\{y_1,y_2,\ldots,y_{n+1}\}$, let it correspond to a binary sequence of length n $x=\{x_1,x_2,\ldots,x_n\}$ such that

$$x_i\equiv y_i+y_{i+1}\pmod 2,\ i=1,2,\ldots,n. \qquad \text{①}$$

Obviously x is determined uniquely by y. On contrary, for each binary sequence of length n $x=\{x_1,x_2,\ldots,x_n\}$ and $y_1=0$ or 1, it corresponds to a binary sequence of length $n+1$ $y=\{y_1,y_2,\ldots,y_{n+1}\}$ such that

$$y_{i+1}=y_i+x_i\pmod 2,\ i=1,2,\ldots,n. \qquad \text{②}$$

Since $y_i + y_{i+1} \equiv y_i + y_i + x_i \equiv x_i \pmod 2$, the correspondence ②
is exactly the inverse correspondence ①. Under the correspondence
①, the four consecutive terms 0, 0, 1, 1 and 1, 1, 0, 0 become the
three consecutive terms 0, 1, 0. On contrary, under the correspondence
②, the three consecutive terms: 0, 1, 0 become the four consecutive
terms: 0, 0, 1, 1 or 1, 1, 0, 0. Therefore each binary sequence of
length n without the three consecutive terms: 0, 1, 0 corresponds just
two binary sequence of length $n + 1$ without the four consecutive
terms: 0, 0, 1, 1 or 1, 1, 0, 0, and we obtain $b_{n+1} = 2a_n$.

Exercise 7

1 Assume that there are n students of the second grade and the
point of each student of the second grade is k. Thus all students get $8 + kn$
points in the aggregate. On the other hand, there are $\binom{n+2}{2}$
tournaments, and the sum of points in each tournament is 1, so
$\binom{n+2}{2} = 8 + kn$, i. e. $n^2 - (2k - 3)n - 14 = 0$. Hence 14 is a multiple
of n. Thus $n = 1, 2, 7$ or 14. Since $n = 1, 2, k < 0$ is impossible and
$n = 7, k = 4$ or $n = 14, k = 8$ are possible, there are 7 students of the
second grade and the points of each student is 4 or there are 14
students of the second grade and the points of each student is 8.

2 Assume that the required smallest value is k. Since the sum of
the interior angles of the quadrilateral equals $360°$ and the sum of the
interior angles of the polygon (or triangle) that obtain after a division
increases at most $360°$. Assume that we obtain $k + 1$ polygons after k
divisions and the sum of the interior angles of these polygons is
S_k, then

$$S_k \leqslant (k + 1) \cdot 360°. \qquad ①$$

On the other hand, the sum of the interior angles of 47 46-sided
polygons is

$$47(46 - 2) \cdot 180° = 47 \times 44 \times 180°$$

and the sum of the interior angles of the remaining $(k + 1) - 47 = k - 46$ polygons (or triangles) is at least $(k - 46) \cdot 180°$, so

$$S_k \geqslant 47 \times 44 \times 360° + (k - 46) \cdot 180°.\qquad\qquad ②$$

From ① and ②, we obtain

$$(k + 1) \cdot 360° \geqslant 47 \times 44 \times 180° + (k - 46) \cdot 180°.$$

Thus $k \geqslant 47 \times 43 - 1 = 2020$. Now we give a division to illustrate that $k = 2020$ is attainable. Firstly, we obtain 47 quadrilaterals after 46 divisions. Secondly, for each quadrilateral, we obtain a 46-sided polygon and 42 triangles after 42 divisions. Thus the total number of divisions is $46 + 47 \times 42 = 47 \times 43 - 1 = 2020$, and we obtain 47 46-sided polygons and 1974 triangles. Therefore the required smallest value of n is 2020.

3 Assume that the conclusion is false, and for any two persons A and B of n persons, among the remaining $n - 2$ persons the number of persons who know one of A, and B is exactly k, then $k \geqslant \dfrac{n}{2}$. Otherwise, among the remaining $n - 2$ persons, there are at least

$$n - 2 - k \geqslant n - 2 - \left(\frac{n}{2} - 1\right) = \frac{n}{2} - 1$$

persons such that each of them are either mutually acquainted or mutually unacquainted with A and B. It contradicts the above assumption.

For any two persons A, B, among the remaining $n - 2$ persons, if there is a person C who knows one of A and B, then $(C; A, B)$ forms a triple and denote the number of such triples by S. On the one hand, there are $\dbinom{n}{2}$ ways to select two persons A, and B from n persons, and there are at least $\dfrac{n}{2}$ triples containing A and B, thus

$$S \geqslant \frac{n}{2}\binom{n}{2}. \qquad \text{①}$$

On the other hand, for any person C_i $(i = 1, 2, \ldots, n)$ of n persons, assume that C_i knows k_i persons in the remaining $n - 1$ persons, thus the number of triples containing C_i is $k_i (n - 1 - k_i)$. So

$$
\begin{aligned}
S &= \sum_{i=1}^{n} k_i (n - 1 - k_i) \\
&= (n - 1) \sum_{i=1}^{n} k_i - \sum_{i=1}^{n} k_i^2 \\
&\leqslant (n - 1) \sum_{i=1}^{n} k_i - \frac{1}{n} \Big(\sum_{i=1}^{n} k_i \Big)^2 \\
&= \frac{1}{n} \Big[\Big(\frac{n(n-1)}{2} \Big)^2 - \Big(\sum_{i=1}^{n} k_i - \frac{n(n-1)}{2} \Big)^2 \Big] \\
&\leqslant \frac{n(n-1)^2}{4} < \frac{n}{2} \binom{n}{2}.
\end{aligned}
$$

It contradicts ①. This completes the proof.

4 Let the set of given points be $I = \{P_1, P_2, \ldots, P_n\}$ and the number of pairs of I with unit distance is E. We draw the circle C_i with the center P_i and unit radius $(i = 1, 2, \ldots, n)$ and assume that there are e_i points of I in the circle C_i $(i = 1, 2, \ldots, n)$. Thus $\sum_{i=1}^{n} e_i = 2E$. The line segment is called a good line segment, if its two end points belong to I. On one hand, there are $\binom{n}{2}$ good line segments.

On the other hand, there are $\binom{e_i}{2}$ chords of the circle C_i which are good line segments, so there are $\sum_{i=1}^{n} \binom{e_i}{2}$ chords of n circle which are all good line segments. But some good line segments as the common chord of two circles are counted repeatedly. Since there are at most $\binom{n}{2}$ common chords of n circles, the number of distinct good line segments is at least $\sum\limits_{i=1}^{n} \binom{e_i}{2} - \binom{n}{2}$. Thus we know

$$\sum_{i=1}^{n} \binom{e_i}{2} - \binom{n}{2} \leqslant \binom{n}{2}.$$

By Cauchy's inequality, we obtain

$$2\binom{n}{2} \geqslant \frac{1}{2}\left(\sum_{i=1}^{n} e_i^2 - \sum_{i=1}^{n} e_i\right)$$

$$\geqslant \frac{1}{2}\left[\frac{1}{n}\left(\sum_{i=1}^{n} e_i\right)^2 - \sum_{i=1}^{n} e_i\right]$$

$$= \frac{1}{2n}(4E^2 - 2nE),$$

i.e. $2E^2 - nE - n^2(n-1) \leqslant 0$, so

$$E \leqslant \frac{n + n\sqrt{8n-7}}{4} \leqslant \frac{n + n\sqrt{8n}}{4} = \frac{n}{4} + \frac{\sqrt{2}}{2}n^{\frac{3}{2}}.$$

5 Let $A = \{a_1, a_2, \ldots, a_n\}$, $P = \{P_1, P_2, \ldots, P_n\}$. Assume that the element a_i of A is regarded as a point in the plane $(i = 1, 2, \ldots, n)$ and the pair $\{a_i, a_j\}$ of elements in A as the line segment joining two points a_i and a_j $(1 \leqslant i < j \leqslant n)$. Thus the problem is equivalent to proving that each point of A is a common end point of just two line segments of P. Assume that the number of line segments starting with a_i is $d_i (i = 1, 2, \ldots, n)$, then

$$\sum_{i=1}^{n} d_i = 2 \mid P \mid = 2n. \qquad \textcircled{1}$$

If the point a_k of A is a common end point of two line segment P_i and P_j of P, then $\{a_k; P_i, P_j\}$ is regarded as a triple, and let the number of such triples be S. Thus $S = \sum\limits_{i=1}^{n} \binom{d_i}{2}$. On the other hand, a_i and a_j are joined by a line segment if and only if P_i and P_j have a common end point a_k forming a triple $\{a_k; P_i, P_j\}$. Hence the number of triple $\{a_k; P_i, P_j\}$ equals the number of line segments joining the points of A, i. e. $S = n$. Thus

$$n = \sum_{i=1}^{n} \binom{d_i}{2} = \frac{1}{2}\left(\sum_{i=1}^{n} d_i^2 - \sum_{i=1}^{n} d_i\right)$$

$$\geq \frac{1}{2}\left[\frac{1}{n}\left(\sum_{i=1}^{n} d_i\right)^2 - \sum_{i=1}^{n} d_i\right]$$

$$= \frac{1}{2n}[4n^2 - 2n^2] = n,$$

i. e. the equality in this inequality holds. By the necessary and sufficient condition that yield equality in Cauchy's inequality hold, we obtain $d_1 = d_2 = \cdots = d_n$. Combining the equality with ①, we obtain $d_1 = d_2 = \cdots = d_n = 2$. It means that each point of A is a common end point of only two line segments of P.

6 Proof I   On one hand, the coefficient of x^n in $(1 + x)^{2n} = \sum_{i=0}^{2n} \binom{2n}{i} x^i$ is $\binom{2n}{n}$.

On the other hand,

$$(1 + x)^{2n} = (1 + 2x + x^2)^n = \sum_{i=0}^{2n} \binom{n}{n-k}(2x)^{n-k}(1 + x^2)^k.$$

Since there exists a team x^n in

$$\binom{n}{n-k}(2x)^{n-k}(1 + x^2)^k = \binom{n}{n-k}(2x)^{n-k}\sum_{i=1}^{k} \binom{k}{i} x^{2i},$$

if and only if k is even and when $k = 2i$, the coefficient of x^n is $\binom{n}{n-2i} 2^{n-2i}\binom{2i}{i}$ in $\binom{n}{n-k}(2x)^{n-k}(1 + x^2)^k$ and $0 \leqslant i \leqslant \left[\frac{n}{2}\right]$. So the coefficient of x^n in $(1 + x)^{2n}$ is also

$$\sum_{i=0}^{[n/2]} \binom{n}{n-2i} 2^{n-2i}\binom{2i}{i}.$$

Hence we obtain

$$\sum_{i=0}^{[n/2]} \binom{n}{n-2i} 2^{n-2i}\binom{2i}{i} = \binom{2n}{n}.$$

Proof Ⅱ   On one hand, the number of ways to select n persons as the deputation from n couples is $\binom{2n}{n}$. On the other hand, we calculate this number in the following way. For any integer i $\left(0 \leqslant i \leqslant \left[\frac{n}{2}\right]\right)$, we select i couples in $\binom{n}{i}$ ways. Afterwords we select $n - 2i$ couples from the remaining $n - i$ couples and select a person from each couple in $\binom{n-i}{n-2i}2^{n-2i}$ ways. Hence for any integer i, we obtain

$$\binom{n}{i}\binom{n-i}{n-2i}2^{n-2i} = \binom{n}{n-2i}\binom{i}{2i}2^{n-2i}$$

ways to select n persons as the deputation and i may equals 0, 1, 2, $\ldots$, $[n/2]$. Therefore there are

$$\sum_{i=0}^{[n/2]}\binom{n}{n-2i}2^{n-2i}\binom{2i}{i}$$

ways to select n persons as the deputation. Hence

$$\sum_{i=0}^{[n/2]}\binom{n}{n-2i}2^{n-2i}\binom{2i}{i} = \binom{2n}{n}.$$

Exercise 8

1 Suppose a_n medals be awarded at the kth day, then

$$a_1 = 1 + \frac{1}{7}(m-1) = \frac{1}{7}(m+6),$$

$$a_k = k + \frac{1}{7}(m - a_1 - a_2 - \cdots - a_{k-1} - k), \qquad \text{①}$$

and

$$a_{k+1} = k + 1 - \frac{1}{7}[m - a_1 - a_2 - \cdots - a_k - (k+1)]. \qquad \text{②}$$

②－①:

$$a_{k+1} - a_k = 1 + \frac{1}{7}(-a_k - 1),$$

i.e. $a_{k+1} - 6 = \frac{6}{7}(a_k - 6)$. So

$$a_k - 6 = (a_1 - 6)\left(\frac{6}{7}\right)^{k-1} = \frac{1}{7}\left(\frac{6}{7}\right)^{k-1}(m - 36),$$

and

$$m = \sum_{k=1}^{n} a_k = \frac{1}{7}(m - 36)\sum_{k=1}^{n}\left(\frac{6}{7}\right)^{k-1} + 6n = (m - 36)\left[1 - \left(\frac{6}{7}\right)^n\right] + 6n,$$

i.e. $m = \frac{7^n}{6^{n-1}}(n - 6) + 36$. Since $(7^n, 6^{n-1}) = 1$, $6^{n-1} \mid (n - 6)$. We could prove that when $n > 1$, $6^{n-1} > \mid n - 6 \mid$. Hence we obtain $n = 6$, and $m = 36$. So the competition last for 6 days and 36 medals are awarded.

2 Let the plane be divided into at most a_n parts by n ellipses. Add an ellipse such that the number of the divided parts of the plane by these ellipses attains maximum, thus the $(n + 1)^{\text{th}}$ ellipse intersect each of the given n ellipses at 4 points and the $4n$ intersecting points are distinct. Thus the $(n + 1)^{\text{th}}$ ellipse is divided into $4n$ arcs and every original region is divided into two parts by each arc. Hence we obtain $a_{n+1} = a_n + 4n$ and $a_1 = 2$, and

$$a_n = a_1 + \sum_{k=1}^{n-1}(a_{k+1} - a_k)$$

$$= a_1 + \sum_{k=1}^{n-1}(a_{k+1} - a_k)$$

$$= 2 + \sum_{k=1}^{n-1}4k = 2n^2 - 2n + 2.$$

Especially, $a_{10} = 182$.

3 Let the permutation $\{a_1, a_2, \ldots, a_n\}$ that satisfies the given condition be called a good permutation with length n and denote the number of such good permutations by S_n. Since $n - 1 \mid a_{n-1} - a_n$ and $\mid a_{n-1} - a_n \mid \leqslant n - 1$, $\mid a_{n-1} - a_n \mid = n - 1$. It follows that $(a_{n-1}, a_n) = (1, n)$ or $(a_{n-1}, a_n) = (n, 1)$. If $(a_{n-1}, a_n) = (1, n)$, then $\{a_1, a_2, \ldots, a_{n-1}\}$ is a good permutation with length $n - 1$. If $(a_{n-1}, a_n) = (n, 1)$, then $\{a_1 - 1, a_2 - 1, \ldots, a_{n-1} - 1\}$ is a good permutation with length $n - 1$. Hence each good permutation with length n corresponds to a good permutation with length $n - 1$. This correspondence is a one-to-one correspondence, so $S_n = S_{n-1}$. Obviously $S_2 = 2$ (There are just two good permutations $\{1, 2\}$, $\{2, 1\}$ with length 2), hence for any positive integer $n \geqslant 2$, the number of permutations satisfying the given condition is $S_n = 2$. When $n = 2m$, there are two good permutations with length n as follows:

$$\{m, m + 1, m - 1, m + 2, m - 2, m + 3, \cdots, 2, 2m - 1, 1, 2m\},$$

and

$$\{m + 1, m, m + 2, m - 1, m + 3, m - 2, \cdots, 2m - 1, 2, 2m, 1\}.$$

When $n = 2m + 1$, there are two good permutation with length n as follows:

$$\{m + 1, m, m + 2, m - 1, m + 3, m - 2, \cdots, 2m - 1, 2, 2m, 1, 2m + 1\},$$

and

$$\{m + 1, m + 2, m, m + 3, m - 1, m + 4, \cdots, 2m, 2, 2m + 1, 1\}.$$

4 Set $A_n = f^{(n)}((1))$ and let b_n denote the number of pairs containing two consecutive teams $0, 0$ in A_n, and c_n denote the number of pair containing two consecutive teams $0, 1$ in A_n. With the given condition, we know through the transformation f only the pair $0, 1$ in A_{n-1} could become the pair $0, 0$ in A_n and only the number 1 or the pair $0, 0$ in A_{n-2} could become the pair $0, 1$ in A_{n-1}. With the definition of f, we obtain that there are 2^{n-2} teams in A_{n-2} and half of these teams are 1s, so $b_n = c_{n-1} = 2^{n-3} + b_{n-2}$. We obtain

$$b_n = 2^{n-3} + 2^{n-5} + \cdots + \begin{cases} 2^0 + b_1, & \text{if } n \equiv 1 \pmod{2}, \\ 2^1 + b_2, & \text{if } n \equiv 0 \pmod{2}, \end{cases}$$

where $b_1 = 0$, and $b_2 = 1$. It follows that $b_n = \frac{1}{3}[2^{n-1} + (-1)^n]$.

5 Let X_n be the set of the alternative sequences whose terms are from the set $\{1, 2, \ldots, n\}$. Thus $|X_n| = A(n)$. When $n = 1$, $X_1 = \{\varnothing, \{1\}\}$, then $A(1) = 2$. When $n = 2$, $X_2 = \{\varnothing, \{1\}, \{1, 2\}\}$, then $A(2) = 3$. For any $\{a_1, a_2, \ldots, a_m\} \in X_n$, by the definition, $a_1 = 1$ or $a_1 \geqslant 3$ is an odd. When $a_1 = 1$, $\{a_2 - 1, a_3 - 1, \ldots, a_m - 1\} \in X_{n-1}$. In this case, let

$$\{a_1, a_2, \ldots, a_m\} \xrightarrow{f} \{a_2 - 1, a_3 - 1, \ldots, a_m - 1\},$$

thus f is a bijective mapping from $X_n^{(1)}$ to X_{n-1}, where $X_n^{(1)} \subset X_n$ and the leading team of each sequence in $X_n^{(1)}$ is $a_1 = 1$, and $|X_n^{(1)}| = |X_{n-1}| = A(n-1)$. When the odd $a_1 \geqslant 3$, $\{a_1 - 2, a_2 - 2, \ldots, a_m - 2\} \in X_{n-2}$. In this case, let

$$\{a_1, a_2, \ldots, a_m\} \xrightarrow{f} \{a_1 - 2, a_2 - 2, \ldots, a_m - 2\},$$

thus f is a bijective mapping from $X_n^{(2)}$ to X_{n-2}, where $X_n^{(2)} \subset X_n$ and the leading team of each sequence in $X_n^{(2)}$ is the odd $a_1 \geqslant 3$. Hence

$$|X_n^{(2)}| = |X_{n-2}| = A(n-2),$$

and

$$A(n) = |X_n| = |X_n^{(1)}| + |X_n^{(2)}| = A(n-1) + A(n-2).$$

It is the recurrence relation of the Fibonacci sequence. Since $A(1) = F_2$, and $A(2) = F_3$, $A(n) = F_{n+1}$, specially, $A(20) = F_{21}$. The first 21 terms of the Fibonacci sequence are 1, 2, 3, 5, 8, 13, 21, 34, 55, 89, 144, 233, 377, 610, 987, 1 597, 2 584, 4 181, 6 765, 10 946, and 17 711, So $A(20) = F_{21} = 17\,711$.

6 Let $b_n = 2^n$, $c_n = n^2$, and $a_n = b_n - c_n = 2^n - n^2$. Since

$$b_{n+3} = 2^{n+3} = 8 \cdot 2^n \equiv 2^n = b_n \,(\mathrm{mod}\ 7),$$

and

$$c_{n+7} = (n+7)^2 \equiv n^2 = c_n,$$

$$a_{n+21} = b_{n+21} - c_{n+21} \equiv b_n - c_n = a_n \pmod{7}.$$

But among a_1, a_2, $\ldots$, a_{21}, just 6 terms: $a_2 = 0$, $a_4 = 0$, $a_5 = 7$, $a_6 = 28$, $a_{10} = 924$, and $a_{15} = 32\,534$ are multiples of 7. Since $9\,999 = 476 \times 21 + 3$ and among a_1, a_2 and a_3, just $a_2 = 0$ is a multiple of 7, there are $476 \times 6 + 1 = 2\,857$ positive numbers n less than 10^4 such that $2^n - n^2$ is a multiple of 7.

7   Using a analogous method in Example 7, firstly, we could prove the following conclusion: Let $a_1 = 1$, $a_2 = 2$ and $a_{n+2} = 3a_{n+1} - a_n (n \in \mathbf{N}_+)$, then

$$3a_n a_{n+1} = a_n^2 + a_{n+1}^2 + 1 (n \in \mathbf{N}_+).$$

(We leave the details of the proof to the reader.) Thus we know that there are infinitely many distinct pairs $(a, b) = (a_n, a_{n+1})(n \in \mathbf{N}_+)$ of positive numbers such that $a^2 + b^2 + 1$ is divisible by ab.

8 **Solution I**   Let $a_1 = 1$, $a_2 = 4$, $a_{n+2} = 4a_{n+1} - a_n (n \geqslant 1)$, $b_n = 2a_{n+1}$ and $c_n = a_{n+2}$, then

$$c_n - b_n = a_{n+2} - 2a_{n+1} = 2a_{n+1} - a_n = b_n - a_n,$$

i.e. a_n, b_n and $c_n (a_n < b_n < c_n)$ construct an arithmetic sequence. Then, we prove that

$$a_n c_n + 1 = a_n a_{n+2} + 1 = a_{n+1}^2.$$

Since $a_3 = 4a_2 - a_1 = 15$, then

$$a_1 c_1 + 1 = 1 \times 15 + 1 = 14 = a_2^2.$$

Assume

$$a_{n-1} c_{n-1} + 1 = a_{n-1} a_{n+1} + 1 = a_n^2,$$

then

$$\begin{aligned}
a_n c_n + 1 &= a_n a_{n+2} + 1 \\
&= a_n (4a_{n+1} - a_n) + 1 \\
&= 4a_n a_{n+1} - a_n^2 + 1 \\
&= 4a_n a_{n+1} - (a_{n-1} a_{n+1} + 1) + 1 \\
&= a_{n+1} (4a_n - a_{n-1}) = a_{n+1}^2.
\end{aligned}$$

Thus for all $n \in \mathbf{N}_+$, $a_n c_n + 1 = a_{n+1}^2$ holds, and

$$a_n b_n + 1 = a_n \cdot 2a_{n+1} + (a_{n+1}^2 - a_n a_{n+2})$$
$$= a_{n+1}^2 - a_n (a_{n+2} - 2a_{n+1})$$
$$= a_{n+1}^2 - a_n (2a_{n+1} - a_n)$$
$$= (a_{n+1} - a_n)^2,$$

$b_n c_n + 1 = (a_{n+2} - a_{n+1})^2$. Hence there exist infinitely many triples $(a, b, c,) = (a_n, b_n, c_n)$ such that $ab + 1$, $bc + 1$ and $ca + 1$ are all perfect squares and a_n, b_n, c_n $(a_n < b_n < c_n)$ construct an arithmetic sequence.

Solution II Let $(2 + \sqrt{3})^n = A_n + B_n \sqrt{3}$ (A_n, B_n are two positive integers, $n \in \mathbf{N}_+$), then $(2 - \sqrt{3})^n = A_n - B_n \sqrt{3}$, $A_n^2 - 3B_n^2 = 1$. Let $a = 2B_n - A_n$, $b = 2B_n$ and $c = 2B_n + A_n$, then a, b, c $(a < b < c)$ construct an arithmetic sequence and $ab + 1 = (A_n - B_n)^2$, $ac + 1 = B_n^2$, and $bc + 1 = (A_n + B_n)^2$. Hence there are infinitely many triples

$$(a, b, c) = (2B_n - A_n, 2B_n, 2B_n + A_n)$$

such that $ab + 1$, $bc + 1$ and $ca + 1$ all are perfect squares.

9 Let $AB = a - d$, $BC = a$, and $CA = a + d$ (a, $d \in \mathbf{N}_+$ and $a > d$), and let S denote the area of $\triangle ABC$, h_a the altitude on the side BC in $\triangle ABC$. Thus

$$S = \sqrt{\frac{3a}{2} \left(\frac{a}{2} + d \right) \cdot \frac{a}{2} \cdot \left(\frac{a}{2} - d \right)}$$
$$= \frac{1}{2} a \cdot \sqrt{3 \left[\left(\frac{a}{2} \right)^2 - d^2 \right]}.$$

Since S is an integer, a must be an even. Set $a = 2x$ ($x \in \mathbf{N}_+$), then $S = x \sqrt{3(x^2 - d^2)}$, $h_a = \frac{2s}{a} = \sqrt{3(x^2 - d^2)}$, and $h_a^2 = 3(x^2 - d^2)$, so h_a must be a multiple of 3. Let $h_a = 3y$ ($y \in \mathbf{N}_+$), then $x^2 - 3y^2 = d^2$. For convenience, we take $d = 1$, thus

$$x^2 - 3y^2 = 1, \qquad \qquad ①$$

and $(x, y) = (2, 1)$ is a solution of ①. Let

$$(2 + \sqrt{3})^n = x_n + y_n \sqrt{3} \, (x_n, \, y_n, \, n \in \mathbf{N}_+),$$

then $(2 - \sqrt{3})^n = x_n - y_n \sqrt{3}$ and $x_n^2 - 3y_n^2 = 1$, i. e. $(x, \, y) = (x_n, \, y_n)(n \in \mathbf{N}_+)$ is the positive integer solutions of ①. Since

$$\begin{aligned} x_{n+1} + y_{n+1}\sqrt{3} &= (2 + \sqrt{3})^{n+1} \\ &= (2 + \sqrt{3})(x_n + y_n\sqrt{3}) \\ &= (2x_n + 3y_n) + (x_n + 2y_n)\sqrt{3}, \end{aligned}$$

$$\begin{cases} x_{n+1} = 2x_n + 3y_n, \\ y_{n+1} = x_n = 2y_n, \end{cases} \text{and} \begin{cases} x_1 = 2, \\ y_n = 1. \end{cases}$$

Eliminating y_n, we obtain $x_{n+2} = 4x_{n+1} - x_n$, $x_1 = 2$, $x_2 = 7$. Set $AB = 2x_n - 1$, $BC = 2x_n$, and $CA = 2x_n + 1$, $(n = 1, 2, 3, \ldots)$, then AB, BC, and CA $(AB < BC < CA)$ construct an arithmetic sequence consisting of coprime positive integers and the area of $\triangle ABC$ is

$$S = x \sqrt{3(x^2 - 1)} = 3x_n y_n,$$

and the altitude on the side BC in $\triangle ABC$ is

$$h_a = \frac{2s}{a} = \sqrt{3(x^2 - d^2)} = 3y_n.$$

So the area and the altitude on the side BC in $\triangle ABC$ are two positive integers. Therefore there exist infinitely many $\triangle ABC$ satisfying the given condition in the problem.

Exercise 9

1   As figure, the 11×12 rectangle is colored with black and white such that no two black cells are covered by a 1×7 or 1×6 rectangle. Since there are twenty black cell in this figure, they are cannot be covered by nineteen 1×7 or 1×6 rectangles. Therefore we have proved that the 11×12 rectangle cannot be covered by nineteen 1×7 or 1×6 rectangles.

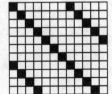

2 (1) Suppose that the $m \times n$ chessboard can be covered by several "L-shapes", then $m \geqslant 2$ and $n \geqslant 2$. Since each "L-shape" contains 4 unit cells, mn is a multiple of 4. Without loss of generality, assume that m is even. Let the n unit cells in the odd rows be colored black and n unit cells in the even rows be colored white. Thus each "L-shape" covers either three white unit cells and a black unit cell or three black unit cells and a white unit cell. In other words, the number of black cells or white cells which are covered by an "L-shape" is odd. If mn is not a multiple of 8, then $mn = 4(2k+1)$ $(k \in \mathbf{N_+})$. If the $m \times n$ chessboard is covered by several "L-shapes", the number of the "L-shapes" must be $2k+1$. But the sum of odds whose number is odd is also odd. It implies that there are odd number of black cells and odd number of white cells in the $m \times n$ chessboard. But the number of black cells and the number of white cells are $2(2k+1)$, which is even, and this is a contradiction. Therefore we have proved that if the $m \times n$ chessboard is covered by several "L-shapes", then mn is a multiple of 8.

(2) From (1), we know that the $m \times n$ chessboard can be covered by several "L-shapes" then mn is a multiple of 8. We will prove that this condition is also sufficient.

(a) if m and n are two evens, then one of m, n is a multiple of 4 since mn is a multiple of 8. Without loss of generality, let $m = 2m_1$, and $n = 4n_1$ $(m_1, n_1 \in \mathbf{N_+})$, then the $m \times n$ chessboard can be divided into $m_1 n_1$ 2×4 rectangles. Since each 2×4 rectangle can be covered by two "L-shapes" (As figure (a)), the $m \times n$ chessboard can be covered by $2m_1 n_1$ "L-shapes".

(b) If one of m and n is odd, then another is a multiple of 8. Without loss of generality, let $m = 2m_1 + 1$, $n = 8n_1$ $(m, n \in \mathbf{N_+})$, then the $m \times n$ chessboard can be divided into two parts, where one part is a $2(m_1 - 1) \times 8n_1$ chessboard, and another part is a $3 \times 8n_1$ chessboard. Since the first part can be divided into $2(m_1 - 1)n_1$ 2×4 rectangles and the second part can be divided into n_1 3×8 rectangles and each 3×8 rectangle can be covered by 6 "L-shapes", (As figure (b)) the $m \times n$ chessboard could be covered by several "L-shapes".

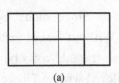   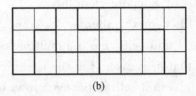

        (a)   (b)

3 We represent n persons with n points. If two persons are acquainted, the two corresponding points are connected by a red line segment, otherwise, connected by a blue line segment. Thus we obtain a 2-colored complete graph K_n and the problem is equivalent to find the smallest positive integer n such that there exist two monochromatic triangles with a common vertex in the 2-colored complete graph K_n. As figure, in the 2-colored complete graph K_8 (the solid lines represent the red lines but we do not draw the blue lines), there is not a blue triangle but 8 red triangles and any two of these 8 red triangles have a common side or do not have a common vertex.

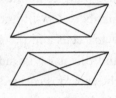

Hence the required smallest positive integer $n \geqslant 9$. Let 9 vertices of the 2 - colored complete graph K_9 be A_1, A_2, ..., and A_9, and the number of red and blue line segments meeting A_1 be x_i and $8 - x_i$ ($i = 1, 2, \ldots, 9$) respectively. Assume that a red line and a blue line meeting some same point form a distinct color angle, thus the number of the distinct color angles of the vertex A_i is $x_i(8 - x_i)$ ($i = 1, 2, \ldots, 9$). If a triangle has red and blue sides, then this triangle is called a distinct color triangle. Since there are two distinct angles in each distinct color triangle, in the 2-colored complete graph K_9, the number of distinct color triangles is

$$\frac{1}{2}\sum_{i=1}^{9} x_i(8 - x_i) \leqslant \frac{1}{2}\sum_{i=1}^{9}\left(\frac{x_i + 8 - x_i}{2}\right)^2 = \frac{1}{2} \times 9 \times 16 = 72.$$

Hence in the 2-colored complete graph K_9, the number of the same color triangles is at least

$$\binom{9}{3} - 72 = 84 - 72 = 12.$$

The 12 triangles have 36 vertices, but there are just 9 points in K_9. By Pigeonhole Principle, there are at least $\left[\dfrac{36-1}{9}\right] + 1 = 4$ triangles with common vertices. Among the 4 triangles, if there are two triangles with a common vertex, then the conclusion holds, otherwise the 4 triangles have a common side, so all sides of the four triangles are colored with the same color. Without loss of generality, assume that $\triangle A_1 A_2 A_3$, $\triangle A_1 A_2 A_4$, $\triangle A_1 A_2 A_5$ and $\triangle A_1 A_2 A_6$ are all red triangles.

Consider the complete graph K_4 with four vertices A_3, A_4, A_5 and A_6, and if one side of K_4, say $A_3 A_4$, is colored red, then $\triangle A_1 A_3 A_4$ and $\triangle A_1 A_2 A_5$ are two red triangles which have a common vertex A_1. Thus the conclusion holds. Otherwise, all sides of this K_4 are blue, and the red triangle $\triangle A_1 A_2 A_3$ and blue triangle $\triangle A_3 A_4 A_5$ are two triangles with a common vertex A_3.

Thus the conclusion also holds. Summing up the above, we conclude that the smallest positive number n is 9.

4 Just as the third problem, this problem is equivalent to finding the smallest positive integer n such that there exist two monochromatic triangles with a common side in the 2-colored complete graph K_n. As figure K_9 (the solid (red) lines and the dashed (blue) lines are constructed in two distinct figures respectively), there are 6 red triangles and 6 blue triangles, and any two monochromatic triangles do not have a common side.

Hence the required smallest positive integer $n \geqslant$ 10. Using an analogous way with the third problem, we could prove that there are at least 20 monochromatic triangles in a 2-colored complete graph K_{10} and these triangles have 60 sides.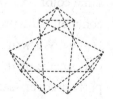

But the number of sides in the K_{10} is $\binom{10}{2} = 45$. By Pigeonhole Principle, there are at least $\left\lceil \frac{60-1}{45} \right\rceil + 1 = 2$ monochromatic triangles with a common side. Summing up the above we obtain that the smallest positive number n is 10.

5   We represent 9 mathematicians with 9 points. If two mathematicians can speak the ith language, the two corresponding points are connected by a line segment with the ith color. Otherwise the line segment connecting the two corresponding points are not colored.

(I) If any two points are connected by a line segment with some color, and the number of colors of line segments meeting each point is at most 3, so by Pigeonhole Principle, there are at least $\left\lceil \frac{8-1}{3} \right\rceil + 1 = 3$ line segments, say A_1A_2, A_1A_3, and A_1A_4, meeting the point A_1 with the same color. So A_1, A_2, and A_3 can speak the same language.

(II) If there exist two points A_1 and A_2 such that the line segment connecting the two points are not colored and with the given condition, we know among any three points, there exist two points such that the line segment connecting the two points are colored with some color, thus for any point $A_i (i = 3, 4, 5, 6, 7, 8, 9)$, there are at least one of A_iA_1 and A_iA_2 is colored with some color. By Pigeonhole Principle, there exists one point of A_1 and A_2, say A_1, such that there are at least $\left\lceil \frac{7-1}{2} \right\rceil + 1 = 4$ line segments meeting A_1 which are colored. But there are just three colors, so there must be two line segments, say A_1A_3 and A_1A_4, which are colored with the same color. So that A_1, A_3, A_4 can speak the same language. This completes the proof.

6   We write 1 in each unit square of the odd rows and -1 in each unit square of the even rows of the 8×8 chessboard. If there exists a perfect cover satisfying the given condition, then the sum of numbers which are covered by a 2×2 square is $2 \times (-1) + 2 \times 1 = 0$ and the sum of numbers which are covered by an "L-shape" is either $3 \times 1 + (-1) = 2$ or $3 \times (-1) + 1 = -2$.

Assume that there are x "L-shapes" and each cover 4 numbers whose sum is 2, and each of the remaining $(9 - x)$ "L-shapes" covers 4 numbers whose sum is -2, then the sum of number which are covered by these figures is $7 \times 0 + 2x + (-2)(9 - x) = 4x - 18$, but the sum of numbers in this 8×8 chessboard is $32 \times 1 + 32 \times (-1) = 0$, so $4x - 18 = 0$, $x = \dfrac{9}{2}$ is not an integer, and this is a contradiction.

Hence there is not a perfect cover satisfying the given condition.

7 When two cogwheels have 14 teeth respectively, assume that each tooth of the cogwheel A (B) is assigned an integer $a_i (b_i)$ $(i = 1, 2, \ldots, 14)$ in clockwise. For the tooth knocked off, let $a_i (b_i) = 1$, otherwise $a_i (b_i) = 0$. We turn the cogwheel A clockwise such the first tooth of A and the ith tooth of B are in gear and let $S_i = \displaystyle\sum_{k=1}^{14} a_k b_{k+i-1}$, $(i = 1, 2, \ldots, 14; b_{j+14} = b_j)$, thus $S_i = 0$ indicates that the projection of two cogwheels is the projection of a whole cogwheel. If for any i $(1 \leqslant i \leqslant 14)$, $S_i \neq 0$, then $S_1 = 4$, $S_j \geqslant 1$. Thus

$$\sum_{i=1}^{14} S_i \geqslant 4 + 13 = 17.$$

But

$$\sum_{i=1}^{14} S_i = \left(\sum_{i=1}^{14} a_i \right) \left(\sum_{i=1}^{14} b_i \right) = 4 \times 4 = 16,$$

and this is a contradiction. Hence there exists an $i_0 (1 \leqslant i_0 \leqslant 14)$ such that $S_{i_0} = 0$. So if we turn the cogwheel A clockwise such that the first tooth of A and the i_0^{th} tooth of B are coincident, then the projection of two cogwheels is the projection of a whole cogwheel. When the two cogwheels have 13 teeth respectively, the conclusion does not hold. In fact, assuming that the teeth knocked off are the first, the second, the fifth and the seventh tooth clockwise and assume that we must turn C_{ij}

tooth clockwise such that the ith tooth of A and the jth tooth of B are coincident. From the following table, we know that whether we turn i $(i = 1, 2, \ldots, 13)$ tooth of A, this case always appears that the tooth knocked off of A and the tooth knocked off of B are coincident, hence the projection of two cogwheels cannot be the projection of a whole cogwheel.

C_{ij} ⟍ B / A	1	2	5	7
1	13	1	4	6
2	12	13	3	5
5	9	10	13	2
7	7	8	11	13

Exercise 10

1 Proof by contradiction. Assume that any two triples have two common members or do not have a common member. If two triples A and B have common members, then they have exactly two common members: a and b. If triples B and C have two common members, then at least one of a, and b belongs to C. Hence triples A and C have also two common members, and all triples are divided into several classes such that any two triples in the same class have exactly two common members and any two triples in distinct class do not have a common member. Now we prove that the number of triples k in the same class is less than or equal to the number of persons h in these triples. Obviously, $h \geqslant 3$. When $h = 3$, $k = 1$. When $h \geqslant 4$, $k \geqslant 2$, and assume that $\{x, a, b\}$ and $\{y, a, b\}$ are two triples in this class, then other triples in this class just have the following forms: $\{x, y, a\}, \{x, y, b\}$ or $\{a, b, z\}$, where we could choose z in $h - 4$ ways at most, hence $k \leqslant 4 + (h - 4) = h$. Thus the total number of triples is less than or equal to total number of persons, i. e. $n + 1 \leqslant n$, and this is a contradiction. This contradiction implies that the conclusion is true.

2   Each of two cases is impossible. Assume that the rows in this chessboard are numbered with $1, 2, \ldots, 8$ from bottom to top and the columns in this chessboard are numbered with $1, 2, \ldots, 8$ from left to right. Let the square in the ith row and the jth column be assigned with (i, j) and the sum of ordinates of 9 chessmen be S. If the 9 chessmen occupy the 3×3 square in top left (or top right) corner, then S increases $3(6+7+8-1-2-3) = 45$, i.e. the parity of S changes. On the other hand, when some chessman moves once, the sum S increases either 2 (if the direction of the jump is vertical or diagonal) or 0 (if the direction of the jump is horizontal), i.e. the parity of S does not change, so this is a contradiction. Thus we prove that each of the two cases is impossible.

3   For sake of contradiction, assume that for any two permutations b and c ($b \neq c$), $S(b) - S(c)$ is not a multiple of $n\,!$, thus $S(a)$ ranges over a perfect complete system of residues with modulus $n\,!$, if a ranges over all $n!$ permutations of $1, 2, \ldots, n$. Let σ denote the set of all permutations of $1, 2, \ldots, n$, thus

$$\sum_{a \in \sigma} S(a) \equiv \sum_{k=1}^{n!} k = \frac{1}{2} \cdot n!(n! + 1) \pmod{n!}.$$

Since $n > 1$ is an odd, $\displaystyle\sum_{a \in \sigma} S(a) \equiv \frac{1}{2} \cdot n! \pmod{n!}$, i.e. $\displaystyle\sum_{a \in \sigma} S(a)$ is not a multiple of $n!$, which contradicts

$$\sum_{a \in \sigma} S(a) \equiv \sum_{a \in \sigma} \sum_{i=1}^{n} k_i a_i = \sum_{i=1}^{n} k_i \Big(\sum_{a \in \sigma} a_i \Big) = \frac{n! \, (n+1)}{2} \times \sum_{i=1}^{n} k_i$$
$$\equiv 0 \pmod{n!}.$$

Hence the conclusion holds.

4   We use 17 points to represent 17 persons. Assume that two points are connected by a line segment if and only if two corresponding persons are mutually acquainted, thus we obtain a graph G with 17 vertices. If two persons are acquainted (i.e.

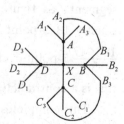

the two corresponding points are connected by a line segment) or have a common acquaintance (i. e. the two corresponding points spanning an angle) then we say that the two corresponding points are associated. Thus the problem is equivalent to proving there exist two points which are not associated.

For the sake of contradiction, assume that any two points are associated, i. e. for any two points they are connected by a line segment or spanning an angle. There are $\frac{1}{2} \times 17 \times 4 = 34$ line segments and $17\binom{4}{2} = 102$ angles and $34 + 102 = 136 = \binom{17}{2}$, which equals the number of two-points groups of 17 points.

Hence for any two points, they are connected by a line segment or construct a unique angle, and one of the two cases occurs. Thus neither a triangle nor a quadrangle occurs in the graph G. Consider any point X, then the point X and four points A, B, C, D are connected by four line segments XA, XB, XC, XD. Thus any two of A, B, C, D are not connected by a line segment and each of A, B, C, D and other three points, except X, A, B, C and D, are connected by line segments respectively. At this time, there are 16 line segments and 4 three-point groups: $\{A_1, A_2, A_3\}, \{B_1, B_2, B_3\}$, $\{C_1, C_2, C_3\}, \{D_1, D_2, D_3\}$ in the figure (as figure). Hence we need to connect $34 - 16 = 18$ line segments.

Since any two points of a three-point group are not connected by a line segment, it is allowed that each point and any point of other three-point groups are connected by a line segment. Thus we obtain a 5 - point cycle containing the point X, as long as we connect a line segment (as figure: XAA_2B_1BX and XBB_3C_3CX) and there are exactly 18 5-point cycles containing the point X. Hence there are $\frac{18 \times 17}{5}$ 5-point cycles in this figure, but $\frac{18 \times 17}{5}$ is not an integer, and this is a contradiction.

This completes the proof.

5 Assume that there exist n persons satisfying the given conditions, and let the 8 acquaintances of some person x be x_1, x_2, ..., x_8. Since x and x_8 are acquainted, then they have 4 common acquaintances. Without loss of generality, assume that the common acquaintances of x and x_8 are x_1, x_2, x_3, x_4, thus x_7 and x_8 are unacquainted. Since x and x_7 are acquainted, then they have 4 common acquaintances which are 4 persons of x_1, x_2, x_3, x_4, x_5, x_6. Hence among the 4 common acquaintances, there are at least 2 persons of x_1, x_2, x_3, x_4. Assume that the 2 persons are x_1 and x_2 Thus x_7 and x_8 are unacquainted, but they have 3 common acquaintances x, x_1, x_2, which contradicts the given conditions. So there is no positive number n satisfying the given conditions.

6 Assume the pair of players playing the ith game is (a_i, b_i) $(i = 1, 2, ..., 14)$ and $M = \{(a_i, b_i) \mid i = 1, 2, ..., 14\}$. A subset S of M is called a good subset, if the players belonging to player pairs of S are distinct. Obviously, good subset exists. (Since one-element subset of M is a good subset) and the number of good subsets is finite. Hence there exists a good subset M_0 such that number of elements of M_0 is the largest.

Assume that there are r pairs of players belonging to M_0. We just need to prove $r \geqslant 6$. If $r \leqslant 5$, then any two of the remaining $20 - 2r$ players, except $2r$ players in M_0, do not play a game. (Since if (a, b) plays a game, then $M_0 \cup \{(a, b)\}$ is also a good subset, which contradicts the definition of M_0.) Thus each of the $20 - 2r$ players plays a game against one of the $2r$ players in M_0.

Hence the number of games at least is $r + (20 - 2r) = 20 - r \geqslant 15$. It contradicts that the total of games is 14. Therefore there exist 6 games such that 12 players who play these games are distinct.

7 Assume that among any three teams, there exist two teams playing a game after S games, and the team A plays k games and k is the smallest. Thus each of teams B_1, B_2, ..., B_k which plays a game against A play at least k games. Let the remaining $19 - k$ teams except A, B_1, B_2, ... , B_k be C_1, C_2, ... , C_{19-k}. Thus for any two teams C_i and C_j ($1 \leqslant i < j \leqslant 19 - k$), there exist two teams of A, C_i and C_j playing a game. But team A and each of teams C_i and C_j do not play a game, hence two teams C_i and C_j play a game, i. e. each of C_1, C_2, ... , C_{19-k} play $18 - k$ games against $18 - k$ other teams.

Hence

$$S \geqslant \frac{1}{2}[k(k+1) + (19-k)(18-k)] = (k-9)^2 + 90 \geqslant 90.$$

On the other hand, if 20 teams are divided into two groups, each group has 10 teams such that any two teams in a group play a game and any two teams in distinct group do not play a game, then the number of games is $2\binom{10}{2} = 90$, and for any three teams, there are two teams belonging to a group which played one game against each other. Summing up the above, we obtain that required smallest number of games is 90.

8 Firstly, we prove that the champion exists. Assume that team A wins k times and k is the largest, then team A is the champion. Otherwise, there exists a team B such that A does not excel B, i. e. B wins A or B wins the k teams which are defeated by A. Thus B wins at least $k + 1$ times, which contradicts that k is the largest. With the same reason, we could prove that for any group of teams, there exists a team which excels the other teams. Assume that there exist two champions A and B and A wins B. All teams are divided into three groups: Each of the first group is defeated by A, and each of the second group wins A, but it is defeated by B, and each of the third group wins A and B.

(1) If the third group is non-empty, then there exists a team C which excels the other teams in the third group and C wins A and B, so C is also a champion.

(2) If the third group is empty, and B is the champion, then the

second group is non-empty. Thus there exists a team C which excels the other teams in the second group and C wins A, and A wins B. So C is also a champion. Summing up the above, we know that there exist at least three champions. Therefore it is impossible that there exist exactly two champions.

Exercise 11

1 Let S_A (S_B) denote the sum of distances from the red points (blue points) to the point A (B). We consider a special case: n blue points are at the left of the midpoint M and n red points are at the right of the midpoint M. In this case, obviously, we obtain $S_A = S_B$. The general case is that at least one red point C is at the left of M and at least blue point D is at the right of M. We take a red point C at the left of M and a blue point D at the left of M. If we swap C with D, then the sum of distances from the red points to point A is $S_A' = S_A + CD$ and the sum of distances from the blue points to the point B is $S_B' = S_B + CD$. Thus $S_A' - S_B' = S_A - S_B$ is a constant.

Therefore we could carry out finitely many adjustments such that the above special case occurs, and the conclusion holds.

2 Firstly, we prove that this algebraic sum is odd. Since

$$1 + 2 + \cdots + 2005 = 1003 \times 2005$$

is odd and for any two integer a and b, the parity of $a + b$ and $a - b$ are the same, any algebraic sum of $1, 2, \ldots, 2005$ is odd. Since

$$1 + (2 - 3 - 4 + 5) + (6 - 7 - 8 + 9) + \cdots$$
$$+ (2002 - 2003 - 2004 + 2005) = 1,$$

the required smallest value is 1.

3 Let $n_1, n_2, \ldots, n_{30}$ denote the number of points in 30 groups respectively, then $n_1, n_2, \ldots, n_{30}$ are distinct and $n_1 + n_2 + \cdots + n_{30} = 1989$, and the total of triangles is $S = \sum_{1 \leqslant i < j < k \leqslant 30} n_i n_j n_k$. Since the number of ways to partition the groups is finite, the maximum of S

exists. Without loss of generality, assume that $n_1 < n_2 < \cdots < n_{30}$.

Firstly, we prove that if S reaches the maximum, then the following properties hold: (1) $n_{k+1} - n_k \leqslant 2$ ($k = 1, 2, \ldots, 29$). In fact, if there exists a k_0 such hat $n_{k_0+1} - n_{k_0} \geqslant 3$, ($1 \leqslant k_0 \leqslant 29$).

For convenience, assume $k_0 = 1$, i.e. $n_2 - n_1 \geqslant 3$, and set $n'_1 = n_1 + 1$, $n'_2 = n_2 - 1$, $n'_k = n_k$ ($3 \leqslant k \leqslant 30$), and assume that the total of triangles corresponding to the division of $\{n'_1, n'_2, \ldots, n'_{30}\}$ is S', then from

$$S = n_1 n_2 \sum_{3 \leqslant k \leqslant 30} n_k + (n_1 + n_2) \sum_{3 \leqslant i < j \leqslant 30} n_i n_j + \sum_{3 \leqslant i < j < k \leqslant 30} n_i n_j n_k$$

and $n'_1 + n'_2 = n_1 + n_2$ we obtain that

$$S' - S = (n'_1 n'_2 - n_1 n_2) \sum_{3 \leqslant k \leqslant 30} n_k = [n_2 - (n_1 + 1)] \sum_{3 \leqslant k \leqslant 30} n_k > 0,$$

i. e. $S' > S$, which contradicts that assumption that S reaches the maximum.

(2) There is at most one k_0 such that $n_{k_0+1} - n_{k_0} = 2$. In fact, if there exists $k_1 < k_2$ such that $n_{k_1+1} - n_{k_1} = 2$, $n_{k_2+1} - n_{k_2} = 2$, and set $n'_{k_1} = n_{k_1} + 1$, $n'_{k_2} = n_{k_2} - 1$, $n'_i = n_i$ ($i \neq k_1, k_2$). With the same reason, we obtain that the total S' of triangles corresponding to the division of $\{n'_1, n'_2, \ldots, n'_{30}\}$ is greater than S, so this is a contradiction. Since $\sum_{i=1}^{30} n_i = 1989$, it is impossible that $n_{k+1} - n_k = 1$ ($k = 1, 2, \ldots, 29$) holds. Otherwise $n_1, n_2, \ldots, n_{30}$ construct an arithmetical sequence with the common difference 1, so

$$1989 = \sum_{i=1}^{30} n_i = 15(n_1 + n_{30})$$

is divisible by 15, and it is a contradiction. Hence there is a k_0 such that $n_{k_0+1} - n_{k_0} = 2$, i.e. 30 numbers are $n_1, n_1 + 1, \ldots, n_1 + (k_0 - 1), n_1 + (k_0 + 1), \ldots, n_1 + 30$. Thus

$$1989 + k_0 = \frac{1}{2} \times 30 \times (n_1 + 1 + n_1 + 30) = 15(2n_1 + 31).$$

So $k_0 = 6$, $n_1 = 51$. Therefore we couclude that when the numbers of points in 30 groups are 51, 52, 53, 54, 55, 56, 58, 59, ... , 81, the total number of triangles reaches the maximum.

4 Consider the general case: assume that a_1, a_2, ... , a_{2n} is a permutation of 1, 2, ... , $2n$, and find the largest value of $S = a_1 a_2 + a_3 a_4 + \cdots + a_{2n-1} a_{2n}$, proof by induction on n. When $n = 1$, $S_1 = 1 \times 2$. When $n = 2$,

$$S_2 \leqslant \text{Max}\{1 \times 2 + 3 \times 4, \ 1 \times 3 + 2 \times 4, \ 1 \times 4 + 2 \times 3\} = 1 \times 2 + 3 \times 4.$$

Assume that $n = k$, and

$$S_k \leqslant 1 \times 2 + 3 \times 4 + \cdots + (2k - 1)(2k),$$

then when $n = k + 1$,

$$S_{k+1} = a_1 a_2 + a_3 a_4 + \cdots + a_{2k+1} a_{2k+2}.$$

(1) If $2k + 1$, and $2k + 2$ appear in the same term of the above expression, say

$$a_{2k+1} a_{2k+2} = (2k + 1)(2k + 2),$$

then by the inductive hypothesis, we obtain that

$$\begin{aligned} S_{k+1} &= a_1 a_2 + a_3 a_4 + \cdots + a_{2k-1} a_{2k} + (2k + 1)(2k + 2) \\ &\leqslant 1 \times 2 + 3 \times 4 + \cdots + (2k - 1)(2k) + (2k + 1)(2k + 2); \end{aligned}$$

(2) If $2k + 1$, and $2k + 2$ appear in two distinct terms of the above expression, say $a_{2k-1} = 2k + 1$, $a_{2k+1} = 2k + 2$, then we note that

$$\begin{aligned} &[a_{2k} a_{2k+2} + (2k + 1)(2k + 2)] - [(2k + 1)a_{2k} + (2k + 2)a_{2k+2}] \\ &= [a_{2k} - (2k + 2)][a_{2k+2} - (2k + 1)] > 0, \end{aligned}$$

thus

$$\begin{aligned} S_{k+1} &< a_1 a_2 + a_3 a_4 + \cdots + a_{2k} a_{2k+2} + (2k + 1)(2k + 2) \\ &\leqslant 1 \times 2 + 3 \times 4 + \cdots + (2k - 1)(2k) + (2k + 1)(2k + 2), \end{aligned}$$

and we have proved that for all $n \in \mathbf{N}_+$,

$$S_n \leqslant 1 \times 2 + 3 \times 4 + \cdots + (2n - 1)(2n).$$

Back to the original problem, if there exists a partition satisfying the given requirement, then

$$m^2 \leqslant 1 \times 2 + 3 \times 4 + \cdots + 13 \times 14 = 504,$$

so $m \leqslant [\sqrt{504}] = 22$ and from the following two figures, we know that $m = 22$ can be achieved. Therefore the required the largest value of m is 22.

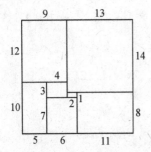 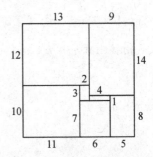

5 Obviously, for any permutation $a = (a_1, a_2, \ldots, a_{20})$ of 1, 2, $\ldots$, 20, we need 19 swaps at most such that the permutation a becomes $(1, 2, \ldots, 20)$, hence $k_a \leqslant 19$. Now we prove that for the permutation $a = (2, 3, \ldots, 20, 1)$, the permutation a could be changed to $(1, 2, \ldots, 20)$, then we need at least 19 swaps. For convenience, we introduce the concept of circuit loop: assume that the number at the b_1^{th} position of the permutation is b_2, and the number at the b_2^{th} position of the permutation is b_3, $\ldots$, and the number at the b_{k-1}^{th} position of the permutation is b_k, and the number at the b_k^{th} position of the permutation is b_1, then

$$b_1 \to b_2 \to \cdots \to b_k \to b_1$$

is called a circuit loop. Especially, if the number at the b_1^{th} position of the permutation is b_1, then $b_1 \to b_1$ is called a circuit loop too. Thus each permutation could be divided into several disjoint circuit loops. Let $f(a)$ denote the number of the circuit loops in the permutation a. We observe the change of $f(a)$ after a swap.

(1) If we swap two numbers b_i, $b_j (i < j)$ in a circuit loop, then

the loop

$$b_1 \to b_2 \to \cdots \to b_k \to b_1$$

is divided into two loops:

$$b_1 \to b_2 \to \cdots \to b_{i-1} \to b_j \to b_{j+1} \to \cdots \to b_k \to b_1$$

and

$$b_i \to b_{i+1} \to \cdots \to b_{j-1} \to b_i.$$

(2) If we swap two numbers b_1 and c_1 in two distinct circuit loops:

$$b_1 \to b_2 \to \cdots \to b_k \to b_1$$

and

$$c_1 \to c_2 \to \cdots \to c_m \to c_1,$$

then this two loops become a single loop:

$$c_1 \to b_2 \to \cdots \to b_k \to b_1 \to c_2 \to \cdots \to c_m \to c_1.$$

Therefore, the value of $f(a)$ either increase 1 or decrease 1 after a swap. Note that $a = (2, 3, \ldots, 20, 1)$ just has a circuit loop: $1 \to 2 \to \cdots \to 20 \to 1$, and $(1, 2, \ldots, 20)$ has 20 circuit loops $(1, 1)$, $(2, 2)$, $\ldots$, $(20, 20)$, thus to obtain $(1, 2, \ldots, 20)$ from $a = (2, 3, \ldots, 20, 1)$, we need at least 19 swaps. Summing up the above, we conclude that the required maximum of k_a is 19.

6　Since there is not a tie, and there is exactly one winner in a game, $\sum_{i=1}^{2n+1} w_i$ equals the total of games, i. e.

$$\sum_{i=1}^{2n+1} w_i = \binom{2n+1}{2} = n(2n+1).$$

By Cauchy's inequality, we obtain

$$n^2(2n+1)^2 = \left(\sum_{i=1}^{2n+1} w_i\right)^2 \leqslant \left(\sum_{i=1}^{2n+1} 1^2\right)\left(\sum_{i=1}^{2n+1} w_i^2\right) = (2n+1)S,$$

so $S \geqslant n^2(2n+1)$. On the other hand, assume that when $i+j$ $(i<j)$ is an odd, p_i wins p_j, and when $i+j$ $(i<j)$ is even, p_j wins p_i, then

for any i $(1 \leqslant i \leqslant 2n+1)$, $w_i = n$. Thus $S = n^2(2n+1)$. Hence the required smallest value of S is $n^2(2n+1)$. To find the largest of S, assume that $w_1 \geqslant w_2 \geqslant \cdots \geqslant w_{2n+1}$. If $w_1 < 2n$, i.e. there are some p_k who wins p_1, and set $w'_1 = w_1 + 1$, $w'_k = w_k - 1$. $w'_i = w_i (i \neq i, k, 1 \leqslant i \leqslant 2n+1)$ and $S' = \sum_{i=1}^{2n+1} w'^2_i$, Thus

$$S' - S = (w_1 + 1)^2 + (w_k - 1)^2 - w_1^2 - w_k^2 = 2(w_1 - w_k) + 2 > 0.$$

When $w'_1 < 2n$ holds, we carry out the same adjustment. Since S is strictly increasing after a adjustment, we obtain $w_1 = 2n$ after finitely many adjustments. Now, we rearrange the $2n+1$ numbers w_1, w_2, ..., w_{2n+1} such that $w_1 \geqslant w_2 \geqslant \cdots \geqslant w_{2n+1}$. From $w_1 = 2n$, we know $w_2 \leqslant 2n-1$. When $w_2 < 2n-1$ holds, we carry out the same adjustment. Thus we obtain $w_2 = 2n-1$ eventually. Then, the adjustment is similar. Thus we obtain

$$S \leqslant (2n)^2 + (2n-1)^2 + \cdots + 2^2 + 1^2 + 0^2 = \frac{1}{3}n(2n+1)(4n+1).$$

On the other hand, assume that for any $1 \leqslant i < j \leqslant 2n+1$, p_i wins p_j, then $w_i = 2n+1-i$ $(1 \leqslant i \leqslant 2n+1))$. In this case,

$$\sum_{i=1}^{2n+1} w_i^2 = \frac{1}{3}n \times (2n+1)(4n+1)$$

holds. Therefore the maximum of $S = \sum_{i=1}^{2n+1} w_i^2$ is $\frac{1}{3}n(2n+1)(4n+1)$.

Exercise 12

1 Assume that $m \times n$ numbers $a_i^2 b_j^2 (i = 1, 2, \ldots, n; j = 1, 2, \ldots, m)$ are filled in each grid of the $m \times n$ table such that each grid contains a number (as figure). The sums of all numbers in each row and each column are perfect squares, as long as $\sum_{i=1}^{n} a_i^2$ and $\sum_{i=1}^{m} b_i^2$ are perfect squares. Note that $(k+1)^2 = (2k+1) + k^2$, we assume that a_1 is odd and $a_2, a_3, \ldots, a_{n-1}$ are even and $0 < a_1 < a_2 < \cdots < a_{n-1}$. Let

$$\sum_{i=1}^{n-1} a_i^2 = 2M + 1(M \in \mathbf{N}_+) \,,$$

and $a_n = M$, then $\sum_{i=1}^{n} a_i^2 = (M + 1)^2$. Then to make these numbers distinct we increase the difference of any two numbers in each row. For example, we set $b_1 > 2a_n$ and b_1 is odd, and $b_2, b_3, \ldots, b_{n-1}$ are even and $b_{i+1} > b_i a_n (i = 1, 2, \ldots, m - 2)$. Denoting $\sum_{i=1}^{m-1} b_i^2 = 2S + 1$ and setting $b_m = S$, we obtain $\sum_{i=1}^{m-1} b_i^2 = (S + 1)^2$. Thus we easily prove that the above way of filling numbers satisfies the given condition of the problem.

$a_1^2 b_1^2$	$a_2^2 b_1^2$	$\cdots$	$a_j^2 b_1^2$	$\cdots$	$a_{n-1}^2 b_1^2$	$a_n^2 b_1^2$
$a_1^2 b_2^2$	$a_2^2 b_2^2$	$\cdots$	$a_j^2 b_2^2$	$\cdots$	$a_{n-1}^2 b_2^2$	$a_n^2 b_2^2$
$\cdots$	$\cdots$	$\cdots$	$\cdots$	$\cdots$	$\cdots$	$\cdots$
$a_1^2 b_i^2$	$a_2^2 b_i^2$	$\cdots$	$a_j^2 b_i^2$	$\cdots$	$a_{n-1}^2 b_i^2$	$a_n^2 b_i^2$
$\cdots$	$\cdots$	$\cdots$	$\cdots$	$\cdots$	$\cdots$	$\cdots$
$a_1^2 b_{m-1}^2$	$a_2^2 b_{m-1}^2$	$\cdots$	$a_j^2 b_{m-1}^2$	$\cdots$	$a_{n-1}^2 b_{m-1}^2$	$a_n^2 b_{m-1}^2$
$a_1^2 b_m^2$	$a_2^2 b_m^2$	$\cdots$	$a_j^2 b_m^2$	$\cdots$	$a_{n-1}^2 b_m^2$	$a_n^2 b_m^2$

2 Let $P = (0, 2ab)$, $Q = (a^2 - b^2, 0)$ (a, b are two integers), then

$$|PQ| = \sqrt{(a^2 - b^2)^2 + (2ab)^2} = a^2 + b^2$$

is an integer. Take the point $P_0(0, 2m)$ in the y-axis, where $m = p_1 p_2 \ldots p_n$ ($p_1, p_2, \ldots, p_n$ are distinct primes) and $Q_i(a_i^2 - b_i^2, 0)$ in the x-axis, where $a_i = p_1 p_2 \ldots p_i$, $b_i = p_{i+1} p_{i+2} \ldots p_n (i = 1, 2, \ldots, n - 1)$. Obviously, n points $P_0, Q_1, Q_2, \ldots, Q_{n-1}$ satisfy the condition of the problem.

3 (1) Since 2003×2005 is not divisible by 3 and there are 3 unit squares in an "L-shapes", the 2003×2005 rectangle cannot be divided perfectly into several "L-shapes".

(2) Since the 2×3 and 7×9 rectangle can be divided perfectly into several "L-shapes". (as figure), and

$$2005 \times 2007 = 1998 \times 2007 + 7 \times 2007$$
$$= (2 \times 3)(999 \times 669) + (7 \times 9) \times 223,$$

the 2005×2007 rectangle can be divided perfectly into several "L-shapes".

4 (1) Assume that the lengths of 8 line segments are 1, 2, 3, 4, 5, 6, 7, 8, respectively in the figure (a). Since any side is greater than the absolute of the difference of other two sides in a triangle, if a triangle is not an isosceles triangle and the lengths of its three sides are all positive integers, then the length of any side of this triangle is greater than 1. Hence the line segment with length 1 is AB or AF. Without loss of generality, assume $AB = 1$. From $1 = AB > | BE - AE |$, we know $BE = AE$. By the cosine theorem in $\triangle ABE$ and $\triangle DEF$, we obtain

$$\cos E = \frac{BE^2 + AE^2 - AB^2}{2BE \cdot AE} = 1 - \frac{1}{2AE^2},$$

and

$$DF^2 = DE^2 + EF^2 - 2DE \cdot EF \cos E$$
$$= DE^2 + EF^2 - 2DE \cdot EF \left(1 - \frac{1}{2AE^2}\right)$$
$$= DE^2 + EF^2 - 2DE \cdot EF + \frac{EF}{AE} \cdot \frac{DE}{BE}.$$

But $\dfrac{EF}{AE}$ and $\dfrac{DE}{BE}$ are two positive fractions less than 1 and the other terms in the above expression are all positive integers, and this is a contradiction. Hence there does not exist 4 straight lines that satisfy the condition of problem.

(2) We start from the rectangular triangle with three sides 3, 4, 5, to find positive integers $k > 1$, m, and n such that the figure (b) satisfies the condition of the problem. By the property of the similar

triangles, we obtain

$$(5 + 4k) : m : (n + 3) = (4 + 5k) : n : (m + 3k) = 3 : 4 : 5.$$

Thus $5 + 4k$ and $4 + 5k$ are all multiples of 3 and we obtain that $k = 4$, $m = 28$, $n = 32$ is a solution. Therefore there are 4 straight lines satisfying the condition of problem (as figure (c)).

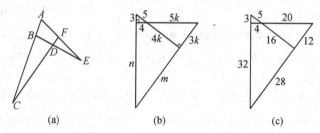

(a)　　　　　　(b)　　　　　　(c)

5 We construct 100 subsets as follows: $A_i = \{k \mid k \equiv i - 1 \pmod{99}, k \in \mathbf{N}_+ \text{ and } k \text{ is even}\}$, where $i = 1, 2, \ldots, 99$, and $A_{100} = \{k \mid k \in \mathbf{N}_+, k \text{ is odd}\}$. Note that if the positive integers a, b, c satisfy $a + 99b = c$, then the number of odd numbers among a, b, c is 2 or 0. If two of a, b, c are odd, then the two numbers belong to A_{100}; If a, b, c are even, and $a \equiv c \pmod{99}$, so a, c belong to the same subset $A_{i_0} (1 \leqslant i_0 \leqslant 99)$.

6 Proof I When $n = 1$, take $A = \{1\}$, $B = \{2\}$, $C = \{3\}$, thus the conclusion holds. Assume that for $n \in \mathbf{N}_+$, the conclusion holds, i. e. the set $S_n = \{1, 2, \ldots, 3n\}$ can be partitioned into three disjoint subsets $A = \{a_1, a_2, \ldots, a_n\}$, $B = \{b_1, b_2, \ldots, b_n\}$, $C = \{c_1, c_2, \ldots, c_n\}$ such that $a_i + b_i = c_i (i = 1, 2, \ldots, n)$, then for $4n \in \mathbf{N}_+$, assume that the set $S_{4n} = \{1, 2, \ldots, 12n\}$ could be partitioned into the following three disjoint subsets: $A' = \{a'_1, a'_2, \ldots, a'_{4n}\}$, $B' = \{b'_1, b'_2, \ldots, b'_{4n}\}$, $C' = \{c'_1, c'_2, \ldots, c'_{4n}\}$, where

$$a'_i = 2i - 1 \ (i = 1, 2, \ldots, 3n), \ a'_{3n+k} = 2a_k (k = 1, 2, \ldots, n),$$
$$b'_i = 9n + 1 - i \ (i = 1, 2, \ldots, 3n), \ b'_{3n+k} = 2b_k (k = 1, 2, \ldots, n),$$
$$c'_i = 9n + i \ (i = 1, 2, \ldots, 3n), \ c'_{3n+k} = 2c_k (k = 1, 2, \ldots, n),$$

then for any $i (1 \leqslant i \leqslant 4n)$, the equality $a'_i + b'_i = c'_i$ holds. Therefore there exist infinitely many positive integers $n = 1, 4, 4^2, 4^3, \ldots$

satisfying the condition of the problem.

Proof II When $n = 1$, take $A = \{1\}$, $B = \{2\}$, $C = \{3\}$, thus the conclusion holds. Assume that for $n \in \mathbf{N}_+$, the conclusion holds, i.e. the set $S_n = \{1, 2, \ldots, 3n\}$ can be partitioned into three disjoint subsets $A = \{a_1, a_2, \ldots, a_n\}$, $B = \{b_1, b_2, \ldots, b_n\}$, $C = \{c_1, c_2, \ldots, c_n\}$ such that $a_i + b_i = c_i$ ($i = 1, 2, \ldots, n$) then for $3n + 1 \in \mathbf{N}_+$, assume that the set $S_{3n+1} = \{1, 2, \ldots 3(3n+1)\}$ could be partitioned into the following three disjoint subsets: $A' = \{a'_1, a'_2, \ldots, a'_{3n+1}\}$, $B' = \{b'_1, b'_2, \ldots, b'_{3n+1}\}$, $C' = \{c'_1, c'_2, \ldots, c'_{3n+1}\}$, where $a'_i = 3a_i - 1$, $b'_i = 3b_i$, $c'_i = 3c_i - 1$ ($i = 1, 2, \ldots, n$) $a'_{n+i} = 3a_i$, $b'_{n+i} = 3b_i + 1$, $c'_{n+i} = 3c_i + 1$ ($i = 1, 2, \ldots, n$), $a'_{2n+i} = 3a_i + 1$, $b'_{2n+i} = 3b_i - 1$, $c'_{2n+i} = 3c_i$ ($i = 1, 2, \ldots, n$), $a'_{3n+1} = 1$, $b'_{3n+1} = 9n + 2$, and $c'_{3n+1} = 9n + 3$, then $a'_i + b'_i = c'_i$ ($i = 1, 2, \ldots, 3n + 1$) holds. Therefore there exist infinitely many positive integers $n = 1, 4, 13, 40, 121, \ldots$, i.e. $n = \frac{1}{2}(3^k - 1)$ ($k = 1, 2, 3, \ldots$) satisfying the condition of the problem.

7 When $n = 1$, let M_1 be a set containing two end points of the line segment with length 1, then M_1 satisfies the condition of the problem. Assume that when $n = k$, there exists a set M_k with finitely many points such that for any $P \in M_k$, there are exactly k points A_1, $A_2, \ldots, A_k \in M_k$ such that the distance between P and A_i equals length 1 ($i = 1, 2, \ldots, k$) and let $|M_k| = m_k$. When $n = k + 1$, we draw m_k circles whose centers are m_k points of M_k respectively and the radius are all equal to 1. Afterward, we connect the center and the intersection point in each circle, and connect any two points of M_k.

Since the number of connecting line segments are finite, there exists a straight line l not parallel to any connecting line segment. We translate the set M_k into the set M'_k along the direction of l such that the distance between any point P of M_k and the corresponding point P' of M'_k is 1.

Thus $M_k \cap M'_k = \varnothing$ and for any $P \in M_k \cup M'_k$, there are exactly $k + 1$ points $P_1, P_2, \ldots, P_{k+1} \in M_k \cap M'_k$ satisfying the distance between P and each of $P_1, P_2, \ldots, P_{k+1}$ is 1, i.e. when $n = k + 1$, the set $M_{k+1} = M_k \cup M'_k$ satisfies the condition of the problem. Hence for any $n \in \mathbf{N}_+$, there exists a set M_n of points satisfying the condition of the problem.

8 We are going to prove a more general result: For any positive integer $m \geqslant 2$, there exists $n_m \in \mathbf{N}_+$ such that when $n \geqslant n_m$, there are m positive integers $a_1, a_2, \ldots, a_m$, satisfying $n = a_1 + a_2 + \cdots + a_m$, $a_1 < a_2 < \cdots < a_m$ and $a_i \mid a_{i+1} (i = 1, 2, \ldots, m-1)$. When $m = 2$, $n = 1 + (n-1)$, then $n_2 = 3$.

Assume that the conclusion is true when $m = k$, i. e. there exists a $n_k \in \mathbf{N}_+$ such that when $n \geqslant n_k$, there are k positive integers b_1, $b_2, \ldots, b_k$ satisfying $n = b_1 + b_2 + \cdots + b_k$, $b_1 < b_2 < \cdots < b_k$ and $b_i \mid b_{i+1}$ $(i = 1, 2, \ldots, k-1)$, then for $n = k + 1$, let $n_{k+1} = 2^{2n_k+1}(2n_k + 1)$.

Thus for any positive integer $n = 2^r(2s + 1) \geqslant n_{k+1} = 2^{2n_k+1} \cdot (2n_k + 1)$, we obtain $2s + 1 \geqslant 2n_k + 1$ or $2^r \geqslant 2^{2n_k+1}$. If $2s + 1 \geqslant 2n_k + 1$, i. e. $s \geqslant n_k$, by the induction hypothesis, we know that there are k positive integers $b_1, b_2, \ldots, b_k$ satisfying $s = b_1 + b_2 + \cdots + b_k$, $b_1 < b_2 < \cdots < b_k$, and $b_i \mid b_{i+1} (i = 1, 2, \ldots, k-1)$.

Then

$$n = 2^r + 2^{r+1}s = 2^r + 2^{r+1}(b_1 + b_2 + \cdots + b_k) = a_1 + a_2 + \cdots + a_{k+1},$$

where $a_1 = 2^r$, and $a_i = 2^{r+1}b_{i-1}(i = 2, 3, \ldots, k+1)$ satisfying $a_1 < a_2 < \cdots < a_{k+1}$ and $a_i \mid a_{i+1}(i = 1, 2, \ldots, k)$. If $2^r \geqslant 2^{2n_k+1}$, i. e. $r \geqslant 2n_k + 1$, then there is $r_1 = 0$ or 1 such that $r = 2p + r_1 (p \in \mathbf{N}_+)$ and from $r = 2p + r_1 \geqslant 2n_k + 1$, we yield $p \geqslant n_k$ and

$$2^p - 1 = (1+1)^p - 1 \geqslant (1+p) - 1 = p \geqslant n_k.$$

By the induction hypothesis, we conclude that there are k positive integers $b_1, b_2, \ldots, b_k$ such that $2^p - 1 = b_1 + b_2 + \cdots + b_k$, $b_1 < b_2 < \cdots < b_k$, $b_i \mid b_{i+1} (i = 1, 2, \ldots, k-1)$. Then

$$n = 2^r(2s + 1) = 2^{r_1}(2s + 1) + (2^r - 2^{r_1})(2s + 1)$$
$$= 2^{r_1}(2s + 1) + 2^{r_1}(2s + 1)(2^p + 1) \times (b_1 + b_2 + \cdots + b_k)$$
$$= a_1 + a_2 + \cdots + a_{k+1},$$

where $a_1 = 2^{r_1}(2s+1)$, and

$$a_i = 2^{r_1}(2s+1)(2^p+1)b_{i-1}(i = 2, 3, \ldots, k+1)$$

satisfying $a_1 < a_2 < \cdots < a_{k+1}$ and $a_i \mid a_{i+1} (i = 1, 2, \ldots, k)$. The proof is complete.

Exercise 13

1   This solution is analogous to Example 1. By the method of enumeration, we could obtain that the answer of (1) to be 174 and answer of (2) to be 78.

2   (1) The number of the regular triangles with the length of side k and its head upward is

$$\begin{aligned}
x_k &= 1 + 2 + \cdots + [n - (k-1)] \\
&= \frac{1}{2}(n-k+1)(n-k+2) \\
&= \frac{1}{6}[(n-k+1) \times (n-k+2)(n-k+3) \\
&\quad - (n-k)(n-k+1)(n-k+2)].
\end{aligned}$$

So the number of the regular triangles with heads upward is

$$S_1 = \sum_{k=1}^{n} x_k = \frac{1}{6}n(n+1)(n+2).$$

The number of the regular triangles whose length of sides is l and head is downward is

$$\begin{aligned}
y_l &= 1 + 2 + \cdots + (n - 2l + 1) \\
&= \frac{1}{2}(n-2l+1)(n-2l+2)\left(1 \leqslant l \leqslant \left[\frac{n}{2}\right]\right).
\end{aligned}$$

When $n = 2m$ is even, the number of the regular triangles whose head is downward is

$$S_2 = \sum_{i=1}^{[n/2]} y_i = \frac{1}{2}\sum_{l=1}^{m}[(2m-2l+2)^2 - (2m-2l+2)]$$

$$= \sum_{l=1}^{m} [2(m-l+1)^2 - (m-l+1)]$$

$$= 2\sum_{l=1}^{m} l^2 - \sum_{l=1}^{m} l$$

$$= 2 \times \frac{1}{6} m(m+1)(2m+1) - \frac{1}{2} m(m+1)$$

$$= \frac{1}{6} m(m+1)(4m-1)$$

$$= \frac{1}{24} n(n+2)(2n-1).$$

When $n = 2m - 1$ is odd, we could obtain

$$S_2 = \sum_{l=1}^{[n/2]} y_l = \frac{1}{2} \sum_{l=1}^{m-1} (2m-2l)(2m-2l+1)$$

$$= 2\sum_{l=1}^{m-1} (m-l)^2 + \sum_{l=1}^{m-1} (m-l)$$

$$= \frac{1}{3} (m-1)m(2m-1) + \frac{1}{2} m(m-1)$$

$$= \frac{1}{6} (m-1)m(4m+1)$$

$$= \frac{1}{24} (n-1)(n+1)(2n+3).$$

Therefore the required number of regular triangles is

$$S = S_1 + S_2 = \begin{cases} \dfrac{1}{8} n(n+2)(2n+1), & \text{if } n \text{ is even,} \\[2ex] \dfrac{1}{8} n(2n^2+3n-1), & \text{if } n \text{ is odd.} \end{cases}$$

(2) Since the bottom half of each rhombus whose sides are not parallel to BC is a regular triangle with head downward. This is a one-to-one correspondence. So the number of the rhombuses whose sides are not parallel to BC equals S_2. Therefore the required number of rhombuses is

$$3S_2 = \begin{cases} \dfrac{1}{8}n(n+2)(2n-1), & \text{if } n \text{ is even,} \\[2ex] \dfrac{1}{8}(n-1)(n+1)(2n+3), & \text{if } n \text{ is odd.} \end{cases}$$

3   Assume that a total of n players participated in the tournament. Thus the n players gathered a total of $2\dbinom{n}{2} = n(n-1)$ points in the tournament. Similarly, the 10 losers get $2\dbinom{10}{2} = 90$ points in games among themselves. Since this accounts for half of their points, they must have a total of 180 points. In games among themselves the $n - 10$ winners similarly gathered

$$2\binom{n-10}{2} = (n-10)(n-11)$$

points. This also accounts for only half of their total number of points (the other half coming from games against the losers), so their total was $2(n-10)(n-11)$ points. Thus we have the equation

$$n(n-1) = 180 + 2(n-10)(n-11),$$

i.e. $n^2 - 41n + 400 = 0$. It follows that $n = 25$ or $n = 16$. But $n = 16$ is rejected, since we must have that $2(n-10)(n-11) \div (n-10) \geqslant 180 \div 10$. Therefore $n = 25$, i.e. there were 25 players in the tournament.

Finally we show that such a tournament exists. Since $n = 25$, we have 15 winners and 10 losers. Every game that the winners play among themselves results in a tie, giving each $15 - 1 = 14$ points from games played with other winners. Likewise, all the games played among the losers results in tie, giving each of the 10 losers 9 points. For the ten games played by each winners against the losers, six are wins, two are losers, and two are ties, giving the winners another 14 points from games played with losers. This gives each loser three wins, nine losses, and three ties in games against winners, adding up to 9 more points.

Thus each of the 25 players receives exactly half of his/her points in games against the losers, which is what we want.

4 Assume that there are m triangles with 3 red sides, n triangles with 2 red sides and 1 blue side, p triangles with 2 blue sides and 1 red side, and q triangles with 3 blue sides. Since the numbers of line segments meeting each point, except point A, are distinct, then they are either $0, 1, \ldots, 16$ or $1, 2, \ldots, 17$. In first case, assume that the number of red line segments meeting the point A is $2k - 1$, then the total of red line segments in the figure equals

$$\frac{1}{2}(0 + 1 + \cdots + 16 + 2k - 1) = 17 \times 4 + k - \frac{1}{2}$$

which is not an integer, and this is a contradiction. Thus the possible case is the second case. Let $B_1, B_2, \ldots, B_{17}$ denote 17 other points except A, and the number of red line segments meeting B_i be i ($i = 1, 2, \ldots, 17$). Thus B_{17} is connected with 17 other points by the red line segments and B_1 is connected with B_{17} by the red line segment. Similarly, for $i = 1, 2, \ldots, 8$, B_{17-i} is connected with $17 - i$ other points except $B_1, B_2, \ldots, B_i$ by the red line segments and B_i is connected with i points $B_{17}, B_{16}, \ldots, B_{18-i}$ by the red line segments. Hence the point A is connected with 9 points $B_{17}, B_{16}, \ldots, B_9$ by the red line segments. The remaining line segments in figure are blue. The angle consisting of two red (blue) line segment meeting the same point is called a red (blue) angle, thus in this figure, the total of red angles is

$$3m + n = \sum_{i=1}^{17} \binom{i}{2} + \binom{9}{2}$$

$$= \binom{3}{3} + \sum_{i=3}^{17} \left[\binom{i+1}{3} - \binom{i}{3} \right] + \binom{9}{2} \qquad ①$$

$$= \binom{18}{3} + \binom{9}{2} = 852.$$

And the total of blue angles is

$$p + 3q = \sum_{k=1}^{15} \binom{17-k}{2} + \binom{8}{2} = \binom{17}{3} + \binom{8}{2} = 708. \qquad ②$$

And the total of red line segments is

$$\frac{1}{16}(3m + 2n + p) = \frac{1}{2}(1 + 2 + \cdots + 17 + 9) = 81. \qquad ③$$

And the total of blue line segments is

$$\frac{1}{16}(n + 2p + 3q) = \frac{1}{2}(16 + 15 + \cdots + 11 + 8) = 72. \qquad ④$$

Combining ①, ②, ③ and ④, we obtain that $m = 204$, $n = 240$, $p = 204$, and $q = 168$, so the number of triangles with three red sides is 204 and the number of triangles with two red sides and one blue side is 240.

5 This solution is analogous to two solutions in the Example 5. We could obtain the answer of the problem is 461.

6 We take the base of the $4 \times 4 \times 4$ cube as the plane of reference. The $4 \times 4 \times 4$ cube is divided into 4 sheaves such that each sheaf is a $4 \times 4 \times 1$ rectangular solid which is parallel to the plane of reference. Let the 4 sheaves are assigned with the numbers 1, 2, 3, 4 (from bottom to top).

We project all red cubes into the plane of reference and each projective small square is assigned with the number of sheaf of this red cube. Thus we obtain a 4×4 table of numbers in which each grid is assigned with one of $\{1, 2, 3, 4\}$ and 4 numbers situated in the same row or same column construct a permutation of $\{1, 2, 3, 4\}$. Such 4×4 table of numbers is called a 4×4 Latin square. On the contrary, each 4×4 Latin square corresponds to a unique coloring way, and this correspondence is bijective.

Hence the problem is equivalent to find the number of 4×4 Latin square. Let a_{ij} denote the number in the unit square at the ith row and the jth column of a 4×4 table ($i, j = 1, 2, 3, 4$). For any permutation $\{a, b, c, d\}$ of $\{1, 2, 3, 4\}$, firstly, we consider the following case: $a_{11} = a$, $a_{12} = a_{21} = b$, $a_{13} = a_{31} = c$, and $a_{14} = a_{41} = d$ (as figure).

a	b	c	d
b	a_{22}	a_{23}	a_{24}
c	a_{32}	a_{33}	a_{34}
d	a_{42}	a_{43}	a_{44}

(1) If $a_{22} = a$, then a_{23} and a_{32} are just equal to d, a_{24} and a_{42} are just equal to c. Thus a_{33} may be equal to a or b, and the values of a_{34}, a_{43}, a_{44} are determined uniquely when the value of a_{33} is determined. Hence in this case, there are 2 Latin squares.

(2) If $a_{22} = c$, then there is only one case: $a_{23} = a_{32} = d$, $a_{24} = a_{42} = a$, $a_{34} = a_{43} = b$ and $a_{33} = a_{44} = a$, Hence in this case, there is 1 Latin square.

(3) If $a_{22} = d$, with the same reason as (2), in this case, there is 1 Latin square. Thus number of Latin squares as figure is exactly 4. The number of such permutations $\{a, b, c, d\}$ of $\{1, 2, 3, 4\}$ is 4!, and if $\{a, b, c, d\}$ is determined, and swap any two of the second row, the third row and the fourth row, we could obtain 3! distinct Latin squares. Therefore the total of 4×4 Latin squares is $4! \times 3! \times 4 = 576$, so the required number of distinct coloring ways is 576.

Remark Let the number of $n \times n$ Latin square be L_n, then $L_n = n!(n-1)!l_n$ and we know that $l_1 = l_2 = l_3 = 1$, $l_4 = 4$, $l_5 = 56$, $l_6 = 9408$, $l_7 = 16942080, \ldots$ When $n \geqslant 5$, the computation of l_n is very complicated, so it is proper that we take $n = 4$ as a problem of the mathematical competition.

Exercise 14

1 Let the colors be a, b, c, d, e, f. Denote by S_1 the sequence a, b, c, d, e, and by S_2 the sequence a, b, c, d, e, f. If $n \in \mathbf{N}_+$ can is represented in the form $5x + 6y$ $(x, y \in \mathbf{N}_+)$, then n satisfies the condition of the problem: we may place x consecutive S_1 sequences followed by y consecutive S_2 sequences around the polygon. Setting y equal to 0, 1, 2, 3 or 4, we find that n may equal any number of the form $5x$, $5x + 6$, $5x + 12$, $5x + 18$, and $5x + 24$. The only numbers greater than 4 not of these forms are 7, 8, 9, 13, 14 and 19.

We prove that none of these numbers has the required property. Assuming a contradiction that a coloring exists for n equal to one of 7, 8, 9, 13, 14 and 19. There exists a number k such that $6k < n < 6(k +$

1). By Pigeonhole Principle, at least $\left[\dfrac{n-1}{6}\right]+1 = k+1$ vertices of the n-sided polygon have the same color. Between any two points of these vertices are at least 4 others, because any 5 consecutive vertices have different colors. Hence, there are at least $5k+5$ vertices, and $n \geqslant 5k+5$. However, this inequality fails when $n = 7, 8, 9, 13, 14$, and 19.

2 Let the colors used be A, B, C, D. We call an unordered pair of squares is sanguine if the two squares lie in the same row and are of distinct colors. Every row gives rise to $6 \cdot 25^2$ sanguine pairs (given by $\dbinom{4}{2}$ possible pairs of colors and 25 squares of each color). Thus summing over all the rows, there is a total of $100 \cdot 6 \cdot 25^2$ sanguine pairs. On the other hand, each such pair is simply the intersection of one row with a pair of distinct columns. Because there are $\dbinom{100}{2} = \dfrac{100 \times 99}{2}$ pair of columns, some pairs of columns contains at least

$$\left[\frac{100 \times 6 \times 25^2 - 1}{100 \times 99/2}\right] + 1 = 76$$

sanguine pairs. Thus some pairs of the fixed columns form sanguine pairs at least 76 rows. We henceforth ignore all other rows and columns. We may as well assume that only a 76×2 board is four colored, in which each row contains two different colors and no color occurs more than 25 times in each column. For each row, consider the pair of colors it contains.

If the pairs $\{A, B\}$ and $\{C, D\}$ each occur in same row, we are done. Likewise for $\{A, C\}$, $\{B, D\}$ and $\{A, D\}$, $\{B, C\}$. Thus, suppose that at most one pair of colors from each of these three sets occurs. Assume the number of the sanguine pairs containing color A is the largest. We can easily see that we have only the following possibilities:

(1) Either $\{A, B\}$, $\{A, C\}$ and $\{A, D\}$ are the only pairs that

can occur, or (2) $\{A, B\}$, $\{A, C\}$ and $\{B, C\}$ are.

In the first case, each of the 76 rows contains a square of color A, implying that one column has at least $\left[\dfrac{76-1}{2}\right]+1 = 38$ squares of color A, a contradiction.

In the second case, the squares of each column can are colored in at most three colors A, B, C. There can only be 25 squares of each color A, B, C in each column, for a total of at most 150 squares-but there are $2 \times 76 = 152$ squares in total, a contradiction. This completes the proof.

3 For sake of contradiction, assume that the conclusion is false. So for two cities X and Y in Γ, there is no oriented road without leaving the republic Γ connecting them. Let A and B denote the set of cities in Γ satisfying that for any $S \in A$, there is an oriented road from X to S without leaving the republic Γ (assume $X \in A$), otherwise, $S \in B$. Obviously $Y \in B \neq \varnothing$. Thus for any $X_1 \in A$ and $Y_1 \in B$, the road connecting X_1 and Y_1 must start from Y_1 to X_1, otherwise, there is an oriented road from X_1 to Y_1, and $X \to X_1 \to Y_1 \in B$, so it is a contradiction.

Let $|A| = a$, $|B| = b$, then $a + b = 668$. If $a \geqslant b$, then $a \geqslant 334 \geqslant b$. Since there are $\dbinom{b}{2}$ oriented roads connecting the cities of B, there is a city $Z \in B$ such that at least $\dfrac{1}{b}\dbinom{b}{2} = \dfrac{1}{2}(b-1)$ oriented roads from Z to other cities of B. But there are a oriented roads from Z to the cities of A. Thus the number of oriented roads starting from city Z is at least

$$a + \frac{1}{2}(b-1) = \frac{a + (a+b) - 1}{2} \geqslant \frac{334 + 668 - 1}{2} > 500,$$

and it contradicts the given condition. If $a < b$, then $a < 334 < b$. Since there are $\dbinom{a}{2}$ oriented roads connecting the cities of A, there is a city $W \in A$ such that at least $\dfrac{1}{a}\dbinom{a}{2} = \dfrac{1}{2}(a-1)$ oriented roads from other cities of A to W, and there are b oriented roads from the cities of B to W. Thus the number of oriented roads to the city W is at least

$$b + \frac{1}{2}(a-1) = \frac{b+(a+b)-1}{2} > \frac{334 + 668 - 1}{2} > 500,$$

and it also contradicts the given condition. Thus the conclusion holds.

4 Let $k = \min\{|A_i|, |B_j|, 1 \leqslant i, j \leqslant n\}$, and without loss of generality, let $|A_1| = k$. Since $B_1, B_2, \ldots, B_n$ are disjoint, there are at most k B_j such that $A_1 \cap B_j \neq \varnothing$, and there are at least $n - k$ B_j such that $A_1 \cap B_j = \varnothing$. Assume that among $B_1, B_2, \ldots, B_n$, there are m sets which are not disjoint with A_1, then $m \leqslant k$. By the given condition: $|B_j \cup A_1| \geqslant n$ and the definition of k, we obtain

$$|M| = |\bigcup_{j=1}^{n} B_j| \geqslant mk + (n-m)(n-|A_1|)$$
$$= mk + (n-m)(n-k) = n(n-k) - m(n-2k).$$

If $n \geqslant 2k$, then

$$|M| \geqslant n(n-k) - k(n-2k) = \frac{n^2}{2} + 2\left(\frac{n}{2} - k\right)^2 \geqslant \frac{n^2}{2}.$$

If $n < 2k$, then $|M| = |\bigcup_{i=1}^{n} A_i| = \sum_{i=1}^{n} |A_i| \geqslant nk > \dfrac{n^2}{2}.$

5 Assume that there are n persons $A_1, A_2, \ldots, A_n$ and 28 problems $T_1, T_2, \ldots, T_{28}$ in this competition, while m problems $T_1, T_2, \ldots, T_m$ are the problems of first test. If A_k has solved two problems T_i and T_j, then $\{A_k; T_i, T_j\}$ forms a triple and assume that the set of all triples is S. By the given condition, we obtain $|S| = n\dbinom{7}{2} = 2\dbinom{28}{2}$, thus $n = 36$.

Next, for any T_i, assume that there are r persons, say $A_1, A_2, \ldots, A_r$, who have solved T_i, thus each of $A_1, A_2, \ldots, A_r$ has also solved other 6 problems except T_i, and there are $6r$ triples of S containing T_i. On the other hand, for any $T_j (j \neq i, 1 \leqslant j \leqslant 28)$, there are exactly two persons who have solved the two problems T_i and T_j. (Since the two persons have solved T_i, they belong to $\{A_1, A_2, \ldots, A_r\}$.) Thus there are 2×27 triples containing T_i, and $6r = 2 \times 27$, i.e. $r = 9$ and it means that each problem is solved by 9 persons.

Assume the conclusion is false, the number of problems solved of each is 1, 2 or 3 in the first test. Let x, y, z denote the number of persons who have solved 1, 2 and 3 problems respectively, then

$$x + y + z = 36. \qquad \text{①}$$

If each person A_i and a problem T_j in the first test which is solved by A_i are paired, then the number of such pairs is

$$x + 2y + 3z = 9m. \qquad \text{②}$$

(Since each problem is solved by just 9 persons.) Moreover, the number of triples containing two first test problems is

$$\binom{2}{2}y + \binom{3}{2}z = 2\binom{m}{2}. \qquad \text{③}$$

From ①, ② and ③, we obtain $x = m^2 - 19m + 108$, $y = -2m^2 + 29m - 108$, and $z = m^2 - 10m + 36$, where $y = -2\left(m - \dfrac{29}{4}\right)^2 - \dfrac{23}{8} < 0$, and it contradicts that y is a nonnegative integer. Thus the conclusion of the problem is true.

6 Let n points be A_1, A_2, $\ldots$, A_n. If two points are connected by a line segment, then this line segment is colored red, otherwise, connected by a blue line segment. Thus we yield a 2-colored complete graph K_n with q red sides and the problem is equivalent to prove that there exist at least $\dfrac{4q}{3n}\left(q - \dfrac{n^2}{4}\right)$ red triangles (i.e. triangles with three red sides).

Assume that there are d_i red line segments meeting A_i and $n - 1 - d_i$ blue line segments meeting A_i ($i = 1, 2, \ldots, n$), then $\sum_{i=1}^{n} d_i = 2q$. Assume that there are α red triangles, β triangles with two red sides and one blue side, and γ triangles with one red side and two blue sides. If an angle consists of two red sides meeting a point, then this angle is called a red angle.

Similarly, if an angle consists of a red and a blue sides meeting a point, then this angle is called a different color angle, thus the number of red angles is

$$3\alpha + \beta = \sum_{i=1}^{n} \binom{d_i}{2}, \qquad \text{②}$$

and the number of different angles is

$$2(\beta + \gamma) = \sum_{i=1}^{n} d_i(n - 1 - d_i). \qquad \text{③}$$

From ③, we know

$$\beta \leqslant \beta + \gamma = \frac{1}{2} \sum_{i=1}^{n} d_i(n - 1 - d_i). \qquad \text{④}$$

Substituting ④ to ②, applying Cauchy's inequality and ①, we obtain

$$\alpha = \frac{1}{3}\left[\sum_{i=1}^{n} \binom{d_i}{2} - \beta\right]$$

$$\geqslant \frac{1}{3}\left[\sum_{I=1}^{N} \frac{1}{2}d_i(d_i - 1) - \frac{1}{2}\sum_{i=1}^{n} d_i(n - 1 - d_i)\right]$$

$$= \frac{1}{3}\left(\sum_{i=1}^{n} d_i^2 - \frac{n}{2}\sum_{i=1}^{n} d_i\right)$$

$$\geqslant \frac{1}{3} \times \left[\frac{1}{n}\left(\sum_{i=1}^{n} d_i\right)^2 - \frac{n}{2}\sum_{i=1}^{n} d_i\right]$$

$$= \frac{1}{3}\left[\frac{1}{n}(2q)^2 - \frac{n}{2}(2q)\right] = \frac{4q}{3n}\left(q - \frac{n^2}{4}\right).$$

This completes the proof.

7 Let $V = \{A_0, A_1, \ldots, A_{n-1}\}$ be the set of all the n vertices, B_i the set of all vertices adjacent to the vertex A_i (i.e. connected with A_i by a line in the figure), and the number of the elements in B_i

denoted by $| B_i | = b_i$. Obviously, $\sum\limits_{i=1}^{n-1} b_i = 2l$ and $b_i \leqslant n-1$, $i = 0$, 1, $2, \ldots, n-1$. If there exists i such that $b_i = n - 1$, without losing generality, we assume that $i = 0$, then the number of lines connecting the $n - 1$ points in B_0 is

$$l - b_0 \geqslant \frac{1}{2} q(q+1)^2 + 1 - (n-1)$$

$$= \frac{1}{2}(q+1)(n-1) + 1 - (n-1)$$

$$\geqslant \frac{3}{2}(n-1) + 1 - (n-1)$$

$$= \frac{n-1}{2} + 1 \geqslant \left[\frac{n-1}{2}\right] + 1.$$

That means there exists a point A_i $(0 \leqslant i \leqslant n - 2)$ which is connected to other two points A_j, $A_k \in B_0$, so there must be a space quadrilateral $A_0 A_j A_i A_k$ in the figure. Next, we consider the case when $b_i \leqslant n - 2$ $(i = 0, 1, 2, \ldots, n-1)$ and we may assume that $q + 2 \leqslant b_0$.

We will give the proof by reduction to absurdity. If there is no such quadrilateral in the figure, B_i and B_j share no vertex-pair when $i \neq j$, then

$$| B_i \cap B_j | \leqslant 1 \ (0 \leqslant i < j \leqslant n - 1).$$

So $| B_i \cap \overline{B_0} | \geqslant b_i - 1$ $(i = 1, 2, \ldots, n-1)$. Then the number of vertex-pairs in $V \cap \overline{B_0}$ is equal to $\binom{n - b_0}{2} \geqslant \sum\limits_{i=1}^{n-1}$ (the number of vertex-pairs in $B_i \cap \overline{B_0}$)

$$\geqslant \sum_{i=1}^{n-1} \binom{| B_i \cap \overline{B_0} |}{2} \geqslant \sum_{i=1}^{n-1} \binom{b_i - 1}{2}$$

$$= \frac{1}{2} \sum_{i=1}^{n-1} (b_i - 1)(b_i - 2)$$

$$= \frac{1}{2} \sum_{i=1}^{n-1} (b_i^2 - 3b_i + 2)$$

$$\geqslant \frac{1}{2}\left[\frac{1}{n-1}\left(\sum_{i=1}^{n-1}b_i\right)^2 - 3\sum_{i=1}^{n-1}b_i + 2(n-1)\right]$$

$$= \frac{1}{2(n-1)}\left[\sum_{i=1}^{n-1}b_i - (n-1)\right] \cdot \left[\sum_{i=1}^{n-1}b_i - 2(n-1)\right]$$

$$= \frac{1}{2(n-1)}(2l - b_0 - n + 1)(2l - b_0 - 2n + 2)$$

$$\geqslant \frac{1}{2(n-1)}(nq - q + 2 - b_0) \cdot (nq - q + 3 - b_0).$$

$\left(\text{Let } \dbinom{b_i - 1}{2} = 0, \text{ if } b_i = 1 \text{ or } 2. \right)$ So

$$(n-1)(n-b_0)(n-b_0-1) \geqslant (nq - q + 2 - b_0)(nq - q - n + 3 - b_0),$$

that is

$$q(q+1)(n-b_0)(n-b_0-1) \geqslant (nq - q + 2 - b_0)(nq - q - n + 3 - b_0). \tag{①}$$

On the other hand

$$(nq - q - n + 3 - b_0) - q(n - b_0 - 1)$$

$$= (q-1)b_0 - n + 3 \geqslant (q-1)(q+2) - n + 3 = 0, \tag{②}$$

and

$$(nq - q + 2 - b_0) - (q+1)(n - b_0)$$

$$= qb_0 - q - n + 2 \geqslant q(q+2) - q - n + 2 = 1 > 0. \tag{③}$$

As $(n - b_0)(q+1)$ and $(n - b_0 - 1)q$ are positive integers, we then have

$$(nq - q + 2 - b_0)(nq - q - n + 3 - b_0) > q(q+1)(n - b_0)(n - b_0 - 1),$$

from ② and ③, which contradicts ①. This completes the proof.

   8   Let

$$A_k = \{k^2 - k + 1, k^2 - k + 2, \ldots, k^2\},$$

$$B_k = \{k^2 + 1, k^2 + 2, \ldots, k^2 + k\}, k = 1, 2, \ldots, n - 1$$

and $S_n = \bigcup\limits_{k=1}^{n-1} A_k$, $T_n = \bigcup\limits_{k=1}^{n-1} B_k$, thus $S_n \cup T_n = X_n$, and $S_n \cap T_n = \varnothing$.
Now, we prove that S_n and T_n are two sets satisfying the condition of

the problem.

When $n = 3$, $S_3 = \{1\} \cup \{3, 4\}$, and $T_3 = \{2\} \cup \{5, 6\}$. Obviously, each of S_3, and T_3 does not contain three elements a_1, a_2, $a_3 (a_1 < a_2 < a_3)$ such that $a_2 \leqslant \dfrac{a_1 + a_3}{2}$. Assume that when $n = k$, each of S_k, and T_k does not contain k elements a_1, a_2, $\ldots$, $a_k (a_1 < a_2 < \cdots < a_k)$ such that $a_i \leqslant \dfrac{a_{i-1} + a_{i+1}}{2}$ $(i = 2, 3, \ldots, k-1)$, i. e. $a_{i-1} - a_i \leqslant a_i - a_{i+1} (i = 1, 2, \ldots, k-1)$.

When $n = k+1$, if the conclusion is false, assume that $S_{k+1} = S_k \cup A_k$ contains $k+1$ elements a_1, a_2, $\ldots$, $a_{k+1} (a_1 < a_2 < \cdots < a_{k+1})$ such that

$$a_i - a_{i-1} \leqslant a_{i+1} - a_i (i = 2, 3, \ldots, k).$$

Thus we know a_k, $a_{k+1} \in A_k$, otherwise, a_1, a_2,$\ldots$, $a_k \in S_k$ $(a_1 < a_2 < \cdots < a_k)$ and $a_i - a_{i-1} \leqslant a_{i+1} - a_i (i = 2, 3, \ldots, k)$, which contradicts the induction hypothesis. Since $|A_k| = k$, then at least one of a_1, a_2, $\ldots$, a_{k-1} belongs to S_k, and assume $a_{i-1} \in S_k$ is the largest $(2 \leqslant i \leqslant k)$. Thus a_i, $a_{i+1} \in A_k$ and $a_{i+1} - a_i \leqslant |A_k| - 1 = k - 1$, $a_i - a_{i-1} \geqslant |B_{k-1}| + 1 = k$, i. e. $a_i - a_{i-1} \geqslant k > a_{i+1} - a_i$, and it contradicts our assumption.

Hence S_{k+1} does not contain $k+1$ elements a_1, a_2, $\ldots$, a_{k+1} satisfying the condition of the problem. With the same reason, we know T_{k+1} does not contain $k+1$ elements a_1, a_2, $\ldots$, $a_{k+1} \in T_{k+1}$ satisfying the condition of the problem. This completes the proof.

Exercise 15

1 Since the differences of any two of 1, 3, 6, 8 are primes, from the given condition, we know $f(1)$, $f(3)$, $f(6)$, $f(8)$ are distinct numbers belonging to A, thus $|A| \geqslant 4$. On the other hand, for any $x \in \mathbf{N}_+$, let $x = 4k + r$, $(k \in \mathbf{N}, r = 0, 1, 2, 3)$, $f(x) = r$ and $A = \{0, 1, 2, 3\}$. Thus for any $x, y \in \mathbf{N}_+$, if $|x - y|$ is a prime and $f(x) =$

$f(y)$, then $4 \mid \mid x - y \mid$, and it contradicts that $\mid x - y \mid$ is the prime. Hence the above function $f : \mathbf{N}_+ \rightarrow A$ satisfies the condition of the problem and $\mid A \mid = 4$, hence 4 is the smallest number of elements in A.

2 $M = \{-3, -2, -1, 0, 1, 2, 3\}$ contains 7 elements, and we claim that it has the required property. Given three distinct elements belonging to M, if some two of them are of opposite sign, then their sum is in M ; if one of them is zero, then the sum of it and either of the other elements is in M. The only other triples of elements of M are $\{-3, -2, -1\}$ and $\{1, 2, 3\}$; in first case, we choose -2 and -1, and in second case, we choose 1 and 2.

Next we claim that 7 is the largest number of elements that M can have. Suppose that there are at least three positive elements belonging to M, and let $b < c$ be the largest two. Given any other positive element $a \in M$, one of sums $a + b$, $a + c$ and $b + c$ must lie in **M**. However, the latter two cannot because they exceed c, the maximal element. Hence $a + b \in M$, and this sum exceeds b, so it must equal c. Therefore $a = c - b$, implying that there is at most one other positive element in M besides b and c. Thus, M cannot have more than three positive elements. Likewise, M cannot have more than three negative elements. M might also contain 0. Therefore 7 is the maximum number of elements that M can have.

3   The total of sets of one-element subsets, two-element subsets and three-element subsets is

$$\binom{20}{1} + \binom{20}{2} + \binom{20}{3} = 1350,$$

and they all satisfy the given condition. Hence the required largest value of n is greater than or equal to 1350. Let $B_1, B_2, \ldots, B_n$ are distinct subsets of M, and when $i \neq j$, $\mid B_i \cap B_j \mid \leqslant 2$. If one of B_1, $B_2, \ldots, B_n$ has at least four elements, without loss of generality, let $\mid B_i \mid \geqslant 4$ and $a \in B_i$, then $(B_i \backslash \{a\}) \cap B_i$ has at least three elements, hence $(B_i \backslash \{a\}) \notin \{B_1, B_2, \ldots, B_n\}$, because any two of B_1, $B_2, \ldots, B_n$ have at most two common elements.

We swap B_i and $B_i \setminus \{a\}$, thus the new family of subsets also satisfies the given condition, and we continue the swap. Eventually, each subset in the family of subsets has at most three elements, and the number of subsets is unchanged. Thus

$$n \leqslant \binom{20}{1} + \binom{20}{2} + \binom{20}{3} = 1350,$$

and the required largest value of n is 1350.

4 If $S = \{1, 2, \ldots, m\}$ does not satisfy the condition of the problem, then there exist two sets A and B ($A \cup B = S$, $A \cap B = \varnothing$) such that for any $n - 1$ numbers (not necessary to be distinct) of A (B), their sum does not belong to A (B). Without loss of generality, let $1 \in A$, then

$$\underbrace{1 + 1 + \cdots + 1}_{n-1} = n - 1 \in B,$$

as long as $m \geqslant n - 1$, and

$$\underbrace{(n-1) + (n-1) + \cdots + (n-1)}_{n-1} = (n-1)^2 \in A,$$

as long as $m \geqslant (n-1)^2$. If $n \in A$, then

$$\underbrace{n + n + \cdots + n}_{n-2} + 1 = (n-1)^2 \in B,$$

and this is a contradiction. If $n \in B$, then

$$\underbrace{n + n + \cdots + n}_{n-2} + (n-1) = n^2 - n + 1 \notin A,$$

as long as $m \geqslant n^2 - n + 1$. But

$$\underbrace{1 + 1 + \cdots + 1}_{n-2} + (n-1)^2 = n^2 - n - 1 \in B,$$

and this is a contradiction. Hence when $m \geqslant n^2 - n - 1$, the set $S = \{1, 2, \ldots, m\}$ has the given property of the problem. Next, let $S_0 = \{1, 2, \ldots, n^2 - n - 2\}$,

$$A = \{1, 2, \ldots, n-2\} \cup \{(n-1)^2, (n-1)^2 + 1, \ldots, n^2 - n - 2\}$$

and

$$B = \{n - 1, n, \ldots, (n - 1)^2 - 1\},$$

then $A \cup B = S_0$, and $A \cap B = \varnothing$. Thus for any $x_1, x_2, \ldots, x_{n-1} \in A$, $x_1 \leqslant x_2 \leqslant \cdots \leqslant x_{n-1}$, if $x_{n-1} \leqslant n - 2$, then

$$n - 1 \leqslant x_1 + x_2 + \cdots + x_{n-1} \leqslant n(n - 2) < (n - 1)^2,$$

i.e. $x_1 + x_2 + \cdots + x_{n-1} \notin A$. If $x_{n-1} > n - 2$, then $x_{n-1} \geqslant (n - 1)^2$, and

$$x_1 + x_2 + \cdots + x_{n-1} \geqslant \underbrace{1 + 1 + \cdots + 1}_{n-2} + (n - 1)^2 = n^2 - n - 1,$$

and $x_1 + x_2 + \cdots + x_{n-1} \notin A$.

Hence A does not contain n numbers $x_1, x_2, \ldots, x_n$ (not necessary to be distinct) satisfying $x_1 + x_2 + \cdots + x_{n-1} = x_n$. Likewise, we could prove that B does not contain n numbers $x_1, x_2, \ldots, x_n$ (not necessary to be distinct) satisfying $x_1 + x_2 + \cdots + x_{n-1} = x_n$. So the set S_0 cannot satisfy the condition of the problem. Therefore the required smallest value of n is $n^2 - n - 1$.

5 Let 8 players be $p_1, p_2, \ldots, p_8$ and let n two-choice (yes or no) questions be $A_1, A_2, \ldots, A_n$. Construct an $8 \times n$ table of numbers as follows: the number in the ith row and the jth column is

$$x_{ij} = \begin{cases} 1, & \text{if for } A_j, \text{ the answer of } p_i \text{ is yes,} \\ 0, & \text{if for } A_j, \text{ the answer of } p_i \text{ is no,} \end{cases}$$
$$(i = 1, 2, \ldots, 8; j = 1, 2, \ldots, n).$$

Thus the sum of numbers in the ith row is $a_i = \sum_{j=1}^{n} x_{ij}$ which repress that for a_i questions of $A_1, A_2, \ldots, A_n$, the answers of p_i are "yes" and the sum of numbers in the jth column is $b_j = \sum_{i=1}^{8} x_{ij}$ which repress that for questions A_j, the answers of b_j persons are "yes". From the given condition, we know there are exactly four 1s and four 0s in each column, hence $b_j = 4$ and

$$\sum_{i=1}^{8} a_i = \sum_{i=1}^{8} \sum_{j=1}^{n} x_{ij} = \sum_{j=1}^{n} \sum_{i=1}^{8} x_{ij} = \sum_{j=1}^{n} b_j = 4n. \qquad ①$$

Note that, if for any column, all 1s are changed to 0 and all 0s are changed to 1, then the table also has the given property. Hence, without loss of generality, assume that all numbers in the first row are 1s. Thus $a_1 = n$, $\sum_{i=2}^{n} a_i = 3n$. If for the pair of questions (A_j, A_k), the answers of p_i are (yes, yes), then $\{p_i; A_j, A_k\}$ forms a triple. By the given condition, we obtain that the number of triples is

$$\sum_{i=1}^{8} \binom{a_i}{2} = 2\binom{n}{2} = n(n-1).$$

Applying Cauchy's inequality and ①, we yield

$$
\begin{aligned}
n(n-1) &= \sum_{i=1}^{8} \binom{a_i}{2} = \binom{a_1}{2} + \sum_{i=2}^{8} \binom{a_i}{2} \\
&= \binom{n}{2} + \frac{1}{2}\left(\sum_{i=2}^{8} a_i^2 - \sum_{i=2}^{8} a_i\right) \\
&\geqslant \frac{1}{2}n(n-1) + \frac{1}{2}\left[\frac{1}{7}\left(\sum_{i=2}^{8} a_i\right)^2 - \sum_{i=2}^{8} a_i\right] \\
&= \frac{1}{2}n(n-1) + \frac{1}{14}\left[(3n)^2 - 7(3n)\right] \\
&= \frac{2}{7}n(4n-7),
\end{aligned}
$$

so $n \leqslant 7$. Now we give an example to illustrate that $n = 7$ is attainable as follows. Hence the required largest value of n is 7.

	A_1	A_2	A_3	A_4	A_5	A_6	A_7
p_1	1	1	1	1	1	1	1
p_2	1	0	0	0	0	1	1
p_3	1	0	0	1	1	0	0
p_4	1	1	1	0	0	0	0
p_5	0	1	0	1	0	1	0
p_6	0	1	0	0	1	0	1
p_7	0	0	1	1	0	0	1
p_8	0	0	1	0	1	1	0

6 Let $S_i = \{n \mid n \in S \text{ and } n \equiv i \pmod{125}\}$ $(i = 0, 1, 2, \ldots, 124)$, then

$$|S_1| = |S_2| = |S_3| = |S_4| = |S_5| = 17,$$

and

$$|S_0| = |S_6| = |S_7| = \cdots = |S_{124}| = 16.$$

Obviously, there is at most one number of S_0 in A and there exists at most one of sets S_i and S_{125-i} whose elements are all in A. When the number of elements in A attains maximum, all the elements of sets S_1, S_2, S_3, S_4, S_5 must be in A. Hence A contains at most

$$1 + |S_1| + |S_2| + \cdots + |S_{62}| = 1 + 5 \times 17 + 57 \times 16 = 998$$

elements. On the other hand,

$$A = \{125\} \cup S_1 \cup S_2 \cup \cdots \cup S_{62}$$

satisfies the given condition of the problem and $|A| = 998$. Hence the required largest value of k is 998.

7 (1) Since the number of ways of partitioning is finite, then there exists a figure G such that $m(G) = m_0$ holds. Let figure G consist of point sets $X_1, X_2, \ldots, X_{83}$ and $m(G) = m_0$ hold. Let X_i contain x_i points, then $\sum_{i=1}^{83} x_i = 1994$, and $m_0 = \sum_{i=1}^{83} \binom{x_i}{3}$. Using the same way as in Example 7, we could prove that for any $1 \leqslant i < j \leqslant 83$, $|x_i - x_j| \leqslant 1$. Note

$$1994 = 83 \times 24 + 2 = 81 \times 24 + 2 \times 25,$$

and when $m(G) = m_0$ holds, among $x_1, x_2, \ldots, x_{83}$, there are 81 numbers which equal 24 and 2 numbers which equal 25. Therefore

$$m_0 = 81\binom{24}{3} + 2\binom{25}{3} = 168\,544.$$

(2) From (1), we know that G^* consists of 83 sub-figures G_1^*, $G_2^*, \ldots, G_{83}^*$, where each of $G_1^*, G_2^*, \ldots, G_{81}^*$ contains 24 points, and each of G_{82}^*, G_{83}^* contains 25 points and for any $1 \leqslant i < j \leqslant 83$, G_i^*

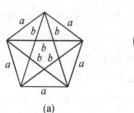

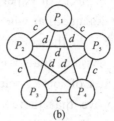

(a)   (b)

and G_j^* are not connected by a line segment. Let point set X_i consist of points of G_i^* ($i = 1, 2, \ldots, 83$). For G_{83}^*, set

$$X_{83} = P_1 \cup P_2 \cup P_3 \cup P_4 \cup P_5, \quad P_i \cap P_j = \varnothing \ (1 \leqslant i < j \leqslant 5),$$

and $\mid P_i \mid = 5$ ($1 \leqslant i \leqslant 5$). Each P_i is colored as figure (a), and the line segments connecting two distinct P_i and P_j are colored as figure (b), where a, b, c, d repress four distinct colors. , thus we obtain a 4-colored graph G_{83}^* which does not contain the triangle whose three sides have the same color, and G_{82}^* is colored in the same way as G_{83}^*.

Finally, for G_i^* ($1 \leqslant i \leqslant 81$), we add a point M to G_i^*, and M and each point in G_i^* is connected by a line segment, then we obtain a new figure $\overline{G_i^*}$. The new figure $\overline{G_i^*}$ is colored in the same way as G_{83}^*, and we delete the point M and line segments meeting M from the new figure $\overline{G_i^*}$. Thus we yield the coloring way of G_i^*. Obviously G_i^* does not contain the triangle whose three sides have the same color. This completes the proof.

8 If we use just three colors, then assume that as figure, among 11 circular pieces of paper, 6 circular pieces of paper are colored with one of the 3 colors, thus each of A, B, C is colored with 1 or 3 color and A, C have the same color but A, C and B have distinct colors. Thus D has a color different from A and B, and D is colored with 2 color. With the same reason, E has a color different from C and B, hence E also is colored with 2 color. However, D and E are circumscribed, and their colors must be distinct, which is a contradiction. It means that just three colors are not enough. Now, we prove that for n circular pieces of paper, four colors are enough. When $n \leqslant$ 4, it is obvious. Assume that when $n = k \geqslant 4$, the conclusion is true, then when $n = k + 1$, the convex hull of centers of $k + 1$ circles is a convex polygon. Assume that A is one vertex of this convex polygon, thus A is the center of some circle.

We could prove that there are at most three circles B, C, D which are circumscribed with the circle A. To move this circle A, by the induction hypothesis, we know each of the remaining k circular pieces of paper can be colored with one of four colors such that any two circles which are circumscribed are colored with distinct colors. Return the circle A, and A could be colored with the fourth color except three colors of B, C, D. Hence for n circular pieces of paper, four colors are enough. Especially, for $n = 2000$, four colors are enough. Therefore the required smallest number of colors is 4.

9 At first, we verify that equations $f(4) = 4$, $f(5) = 5$, and $f(6) = 5$ hold.

When $n = 4$, consider $\{m, m+1, m+2, m+3\}$. If m is odd, then m, $m+1$, $m+2$ are mutually prime. If m is even, then $m+1$, $m+2$, $m+3$ are mutually prime. Thus $f(4) \leqslant 4$. But from the set $\{m, m+1, m+2, m+3\}$, choose a 3-element subset consisting of two evens and one odd, we know these three numbers are not mutually prime, which implies that $f(4) = 4$.

When $n = 5$, consider $\{m, m+1, m+2, m+3, m+4\}$. If m is even, then m, $m+2$, $m+4$ are all evens. Then any 3 numbers from the 4-element subset $\{m, m+1, m+2, m+4\}$ are not mutually prime, so $f(5) > 4$. But from the 5-element universal set we can find out 3 numbers which are mutually prime. Thus $f(5) = 5$.

When $n = 6$, the elements of the set $\{m, m+1, m+2, m+3, m+4, m+5\}$ are 3 odd and 3 even. If we choose a 4-element subset consisting of 3 even and 1 odd, then among the subset there are not 3 numbers which are mutually prime. Thus $f(6) > 4$. Consider the 5-element subset, if among these 5 elements there are 3 odd, then these 3 numbers must be mutually prime. If these 5 elements are 3 even and

2 odd, then among these 3 even, there is at most one number which is divisible by 3, and there is at most one number which is divisible by 5. Hence among these 3 even there is one number which is neither divisible by 3 nor 5. The even and the other 2 odd are mutually prime, which implies that $f(6) = 5$.

When $n > 6$, set $T_n = \{t \mid t \leqslant n + 1 \text{ and } 2 \mid t \text{ or } 3 \mid t\}$. Then T_n is a subset of $\{2, 3, \ldots, n + 1\}$, and any 3 elements in T_n are not mutually prime. By the inclusion-exclusion principle, we have

$$|T_n| = \left[\frac{n+1}{2}\right] + \left[\frac{n+1}{3}\right] - \left[\frac{n+1}{6}\right].$$

So

$$f(n) \geqslant |T_n| + 1 = \left[\frac{n+1}{2}\right] + \left[\frac{n+1}{3}\right] - \left[\frac{n+1}{6}\right] + 1.$$

Assume $n = 6k + r$, $r = 0, 1, 2, 3, 4, 5$, then

$$\left[\frac{n+1}{2}\right] + \left[\frac{n+1}{3}\right] - \left[\frac{n+1}{6}\right] + 1 = 4k + \left[\frac{r+1}{2}\right] + \left[\frac{r+1}{3}\right] - \left[\frac{r+1}{6}\right] + 1.$$

It is easy to verity that

$$\left[\frac{r+1}{2}\right] + \left[\frac{r+1}{3}\right] - \left[\frac{r+1}{6}\right] = \begin{cases} r, & \text{if } r = 0, 1, 2, 3, \\ r - 1, & \text{if } r = 4, 5. \end{cases}$$

If $r = 0, 1, 2, 3$, we can divide $n = 6k + r$ numbers into k groups: $\{m, m + 1, \ldots, m + 5\}$, $\{m + 6, m + 7, \ldots, m + 11\}$, $\ldots$, $\{m + 6(k - 1), m + 6k - 5, \ldots, m + 6k - 1\}$ and left with r numbers $m + 6k, \ldots, m + 6k + r - 1$. Among $4k + r + 1$ numbers, there are at least $4k + 1$ numbers which are contained in the above k groups. Thus there is at least one group which contains 5 numbers. Since $f(6) = 5$, there are 3 numbers which are mutually prime. If $r = 4$, and 5, we can prove similarly that there are 3 numbers which are mutually prime. Hence,

$$f(n) = \left[\frac{n+1}{2}\right] + \left[\frac{n+1}{2}\right] - \left[\frac{n+1}{6}\right] + 1.$$